D0212502

CAMBRIDGE ASTROPHYSICS SERIES

Low light level detectors in astronomy

Low light level detectors in astronomy

M. J. ECCLES

Burr Brown Ltd., Livingston, Scotland

M. ELIZABETH SIM AND K. P. TRITTON

Royal Observatory, Edinburgh

CAMBRIDGE UNIVERSITY PRESS

Cambridge

London New York New Rochelle

Melbourne Sydney

Published by the Press Syndicate of the University of Cambridge
The Pitt Building, Trumpington Street, Cambridge CB2 1RP
32 East 57th Street, New York, NY 10022, USA
296 Beaconsfield Parade, Middle Park, Melbourne 3206, Australia

First published 1983

Printed in Great Britain at the University Press, Cambridge

Library of Congress catalogue card number: 82–12881

British Library Cataloguing in Publication Data

Eccles, M. J.
Low light level detectors in astronomy.
(Cambridge astrophysics series)

1. Imaging systems 2. Image intensifiers 3. Astronomy
I. Title II. Sim, M. Elizabeth III. Tritton, K. P.
621.36'72'024521 TK8316

ISBN 0 521 24088 3

GO

To John Whelan
who encouraged the writing and production of this book
from its beginning

Contents

Preface

This book is based on a course of lectures entitled 'Low Light Level Detectors', given at the University of Edinburgh for the M.Sc. course in Astronomical Technology. Its subject is the comparative study of the light detectors in general use in astronomy. Its scope is limited to the detectors of optical radiation, in particular excluding the group of detectors used only at infrared wavelengths.

After a general introduction, the book falls naturally into three parts: photography (Chapters 2 and 3), electronography (Chapter 4) and the photoelectronic detectors (Chapters 5–8). These parts are essentially independent and could be read in any order. The chapter on electronography was contributed by Dr M. R. S. Hawkins.

No attempt has been made to give detailed references in the text to original work published in journals. Rather, a few key references are given at the end of each chapter, mostly to textbooks, review articles and conference proceedings.

MJE
MES
KPT

1982 August

Acknowledgements

The authors gratefully acknowledge permission to reproduce various figures and photographs. Figures 1.5, 2.7, 2.8, 2.10, 3.1 and 3.2 are reproduced or adapted from Kodak Publications 'Kodak Plates and Films for Scientific Photography', P-315, and 'Kodak Technical Pan Film', P-255, by permission of the Eastman Kodak Company; Figure 2.11 is reproduced with permission of *Sky and Telescope* magazine; Figures 3.11 and 3.12 are printed by permission of the Anglo-Australian Telescope; Figure 7.12 is reproduced with permission from *Astronomical Observations with Television-type Sensors*, edited by J. W. Glaspey and G. A. H. Walker, University of British Columbia, 1973; Figure 7.13 is reproduced by courtesy of the *Publications of the Astronomical Society of the Pacific* volume 91, page 120, 1979 (from Atwood et al.); Figure 8.15, an image taken with the Cambridge CCD camera on the Anglo-Australian Telescope, was supplied by C. D. Mackay.

The cover illustration, an image of NGC 6814 taken with the Cambridge CCD system on the Anglo-Australian Telescope, was supplied by C. D. Mackay.

1
Introduction

When astronomers seek to extend their studies beyond the Solar System, they become almost totally dependent on the detection of quanta of electromagnetic radiation to observe and survey the Universe. The earliest such observations were, naturally enough, made visually, with or without the aid of a telescope. As a detector of light, the eye has many advantages but it suffers from an inability either to store or to quantify the information it receives. It was superseded as an astronomical detector firstly by the photographic emulsion and then more recently by a range of electronic devices that exploit the photoemissive or photoconductive processes. Considerable advances in detector technology have not only enabled astronomers to extend their knowledge by observing fainter and fainter objects, but have also enabled telescopes of only modest sizes to be used in projects that would formerly have required the largest telescopes available.

1.1 Low light level detectors

Stars and galaxies appear as extremely faint light sources, which is why the light gathering power of a telescope must be used to increase their detectability. At the same time, the optical configuration of the telescope is designed to form an image of a region of the sky at a manageable scale and thereby focus it onto the detector being used. This image may be regarded as an input signal, in photons, which is applied to the detector, and the detector may be regarded as a device which converts this input signal into some recordable form. The output signal from the detector may be the blackening of a photographic emulsion, the current from a photomultiplier or television system, or simply numbers stored on a magnetic tape from an electronic counter.

There is no such thing as an ideal detector, though it is possible to draw up a list of the desirable properties that an ideal device would possess. It would be efficient, in the sense that it would be able to record a high proportion of the photons that are incident upon it, and it would be accurate, that is the input signal would be reliably and precisely calculable from the output signal. It would be operable over a wide range of input signals without suffering overload or saturation. It would also be capable of integrating and storing or reading out the output in any desired operational timescale. In the case of detectors designed to record one- or two-dimensional images, the spatial stability, resolution and uniformity would need to be good, and a large detecting area would need to be available. Finally, the detector would be operationally simple, reliable and robust.

The detectors that astronomers actually have at their disposal always fall short of the ideal in one respect or another, and so an appreciation of their strengths and weaknesses is needed in order to ensure their efficient operation.

1.2 Photon arrival rates

Figure 1.1 shows the rate at which photons arrive at the Earth's surface from stars of different apparent magnitudes. The letters U, B, V, R and I stand for various optical wavebands, approximately centred on the wavelengths 360, 440, 550, 700 and 900 nm respectively. Entering the diagram from the bottom with a magnitude in any of these wavebands, a value of log N can be read off vertically above it on the horizontal scale, where N is the number of photons of approximately the stated wavelength that arrive at sea level per square metre per second per nanometre of waveband. An allowance has been made in this diagram for the photons lost by absorption as they pass down through the Earth's atmosphere, assuming that the star is at the zenith. For example, the diagram shows that for a star of magnitude $V = 5$, which is faint but clearly visible to the unaided eye, N is about 8×10^5. In other words about 800 000 photons of wavelength near 550 nm reach each square metre of the Earth's surface from this star each second in each nanometre waveband. Assuming that the eye accepts a useful waveband of 200 nm and taking a pupil area of about 50 mm^2, then the eye should be receiving about 8000 detectable photons from it per second. The eye cannot see stars very much fainter than this, unless aided by the larger collecting area of a telescope.

The diagonal lines in Figure 1.1 represent telescopes at different apertures. Hence the photon collection rate of a given telescope can be obtained by reading from the log N to the log n scale via the appropriate diagonal. Here n is the number of photons collected per second per nanometre of waveband. It can be seen that a 4 m telescope collects over 10^7 photons per second per nanometre from the $V = 5$ magnitude star. Not all of these will reach the focus of the telescope because of reflection and absorption losses from the components in the optical train. For example, at the prime focus of a reflector, with a corrector plate, losses could be about 15 per cent. If the image is being recorded at the telescope focus, then the photon arrival rate n, after correction for losses in the optical train, is the input rate to the detector. If an instrument such as a spectrograph is being used at the focal plane, then the losses in the instrument must also be allowed for. In the case of a slit spectrograph, not all the light enters the instrument, and losses inside it may well be as high as 60 per cent.

The figure shows the enormous range of photon fluxes that must be

Figure 1.1. N is the number of photons per square metre per nanometre per second at the stated wavelength at sea level. n is the number of photons per nanometre per second at the stated wavelength at sea level. For an explanation see the text.

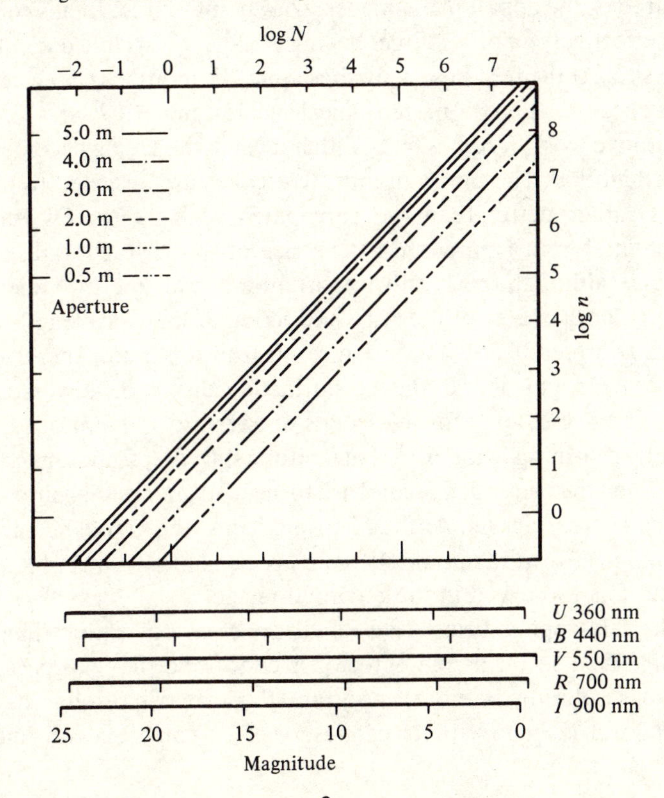

dealt with in optical astronomy. A $B = 24$ magnitude star is just at the limit that a 4 m telescope can detect photographically. Taking a photographic waveband of 100 nm, the whole telescope will intercept from it a mere 35 photons per second.

It is obviously necessary to compare the spatial scale of a detector with the angular scale of the image plane at which it is used. The focal plane scale s, usually expressed as the number of arcseconds in the sky that are imaged on to one millimetre of the focal plane, is determined solely by the focal length F of the focal station. Simple geometry shows that $s = 206\,265/F$ if s is in arcsec mm^{-1} and F is in mm. For example, at an $f/3$ prime focus of a 4 m telescope, $F = 12$ m, and so $s = 17.2$ arcsec mm^{-1}, whereas at the $f/15$ Cassegrain focus, $F = 60$ m and $s = 3.4$ arcsec mm^{-1}. The UK Schmidt telescope has aperture 1.24 m, focal length 3.07 m and a focal plane scale 67.1 arcsec mm^{-1}.

The angular resolution of the image at the focal plane is limited either by the seeing or by the quality of the telescope optics. Seeing is the scintillation of the image caused by turbulence in the atmosphere, and is measured by the apparent angular size of a point source. In this context, a star is effectively a point source. Seeing is rather a variable quantity even at one site. At the best observing sites it may be as little as a few tenths of an arcsecond, though more typical values are 1–2 arcsec. Good quantitative work becomes more difficult when the seeing becomes much worse than this. The optics of most telescopes are designed to produce images rather better than the anticipated seeing disk. Of course, a fundamental limit is imposed by the size of the diffraction disk of the telescope, although this is only important for small aperture telescopes. The angular diameter of the Airy diffraction disk at wavelength λ nm is $0.252\ \lambda/D$ arcsec if D is the aperture measured in mm. Therefore at a visual wavelength of 550 nm the diffraction disk will be worse than a seeing of 1 arcsec only for telescopes of aperture 140 mm or less.

Briefly returning to the $B = 24$ magnitude star, the 35 photons collected by the 4 m telescope each second are focused in good seeing into an area perhaps 1 arcsec across. At the $f/3$ prime focus, this would be an area of only about 60 μm diameter. It depends on the detector whether this resolution is preserved in the recorded image.

If the telescope is being used for slit spectroscopy rather than direct imaging, then so far as the detector is concerned the consideration of spectral resolution is more important than spatial resolution. If the spectrograph is set up so that it can just resolve a pair of spectral lines near wavelength λ that are separated by δλ, then the spectroscopic resolving power is defined as λ/δλ. The linear dispersion determines how far apart these two lines are imaged in the spectrograph camera focal plane. For example, assume a resolving power of 5000 and a dispersion of 5 nm mm^{-1}. At a wavelength of 440 nm, the spectral resolution is 0.088 nm, and this waveband will be dispersed across 17.6 μm of the focal plane. A 2 m telescope would collect about 75 photons s^{-1} in this waveband from a B = 14 magnitude star, though telescope and spectrograph losses could reduce this flux by as much as 90 per cent by the time the light reaches the detector.

At night, the eye cannot see fainter than about $V = 6$ magnitudes because it is not sensitive enough. In daylight, it cannot see stars even this faint, because the sky is too bright. The daytime limit is about $V = -3$ magnitudes, which means that it is possible to see the planet Venus (and, very rarely, the planet Jupiter) in broad daylight, if at its brightest. Obviously the sky brightness affects the detectability of an object whatever the detector, and this point will be elaborated in Section 1.3. The typical brightness of the night sky in different wavebands at one observatory is given in Table 1.1, in which, since the sky is a source of surface brightness, the units are given per square arcsecond of sky. In practice, these figures may vary up or down by a factor of two depending on the time of night and the part of the sky being observed.

All examples given so far have been for ground based telescopes, and would not be appropriate for orbiting or other telescopes operating above the atmosphere. If there is no atmosphere, there is no seeing, and the

Table 1.1. The typical sky background brightness at Mt Palomar Observatory and as observed above the atmosphere, in units of photons $arcsec^{-2}$ m^{-2} s^{-1} nm^{-1}. These figures vary considerably with the direction of observation, being highest in the plane of the Galaxy and the ecliptic. The terrestrial figures also vary with time.

Wavelength (nm)	Mt Palomar Observatory	Above atmosphere
360	0.094	0.049
440	0.14	0.095
550	0.19	0.12
700	0.30	0.14
900	0.91	0.11

image resolution is determined only by the optical quality of the telescope. Consequently the diffraction limit becomes of importance even for larger telescopes. Atmospheric absorption is also absent, allowing access to all wavelengths. The loss from atmospheric absorption is perhaps 20 per cent in the visual waveband, but is total at ultraviolet wavelengths short of 300 nm and apart from certain windows, also substantial in the infrared. Figure 1.2 shows the transmission of the atmosphere to ground level as a function of wavelength.

In space the sky brightness is lower because of the absence of atmospheric scattering of starlight and also because of the absence of emission in the atmosphere itself, but other sources of sky brightness are still present, especially zodiacal light and unresolved starlight. The consequent gain in the detectability of faint objects is not great at optical wavelengths but becomes much better in the infrared. Typical figures for the sky brightness above the atmosphere have been included in Table 1.1 for comparison.

Figure 1.2. The transmission of the atmosphere as measured by an observer at sea level observing a star at the zenith. The absorption bands at the infrared wavelengths vary considerably with the humidity, and are much reduced at high, dry observatories.

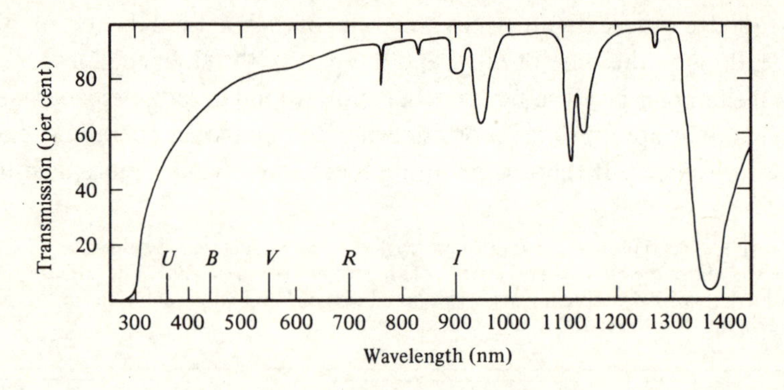

1.3 Photon statistics and noise

Assume that a detector is receiving photons from a star at a rather low arrival rate. The number of photons received in each consecutive time interval is not a constant but fluctuates randomly around some average value. Suppose that in successive intervals of one second, the number of photons received is 3,2,4,1,5,3,2,4,3... and that over a period of 50 seconds the arrivals can be tabulated as follows.

r	0	1	2	3	4	5	6	7	8	>8
Number of seconds in which r photons arrive	2	6	11	11	9	6	3	1	1	0

Counts like these are quite well represented by Poisson statistics, which show that if the mean rate of occurrence of an event is k per unit time, then the probability of r occurrences in any unit time interval is $P(r) = e^{-k}k^r/r!$ It is easy to confirm that this probability distribution has a mean value k, as follows. Over T time units there should be TP units in which there are r events, and so the mean count rate is

$$\frac{1}{T}\sum_{r=0}^{T} rTP = \sum_{r=0}^{T} e^{-k}rk^r/r!$$

As T increases, this expression tends to the value k. The root mean square (rms) deviation of the event rates from the mean value is a measure of the fluctuation in the rate of occurrence. The mean square deviation is

$$\frac{1}{T}\sum_{r=0}^{T} (r-k)^2TP.$$

This expression also tends to the value k as T increases, and the rms deviation tends to the value $k^{1/2}$.

In the example above, the total number of photons arriving in 50 seconds is 160, and so the mean arrival rate is 3.2 photons per second. The mean square deviation of the arrival rate is easily found to be 3.04. According to Poisson statistics, the mean arrival rate and the mean square deviation of the arrival rate should be the same over a long timescale. In the example, the two figures are close enough to each other considering the small sample of time intervals taken. The mean arrival rate of 3.2 photons per second can be regarded as the input signal, and the uncertainty in this quantity after 50 seconds of observation, given by the rms fluctuation, is 1.74 photons per second.

Any source of uncertainty in a signal is called noise, and the fluctuation in the photon arrival rate illustrated above is called photon noise. Noise can be introduced at more than one stage of the detection process and for convenience is here classified broadly into background noise, detector noise and readout noise. Apart from the strongest signals, which are photon noise limited, the final accuracy of the measurement is limited by one or other of these. Some of the more important examples will be noted here.

Background noise is caused by photons reaching the detector from sources other than the object being observed. The most important such source is the sky. Since the number of photons reaching the detector from the sky can be as great as or more than the number from a faint star, the detectability of faint objects may be limited by sky noise. Background noise can also be caused by any other source of stray light. For use at infrared wavelengths the telescope structure itself is a source of radiation and must be shielded from the view of the detector. To be able to subtract out the background contribution to the signal from an object, the signal needs also to be measured in the absence of the object. For example the object and a patch of nearby empty sky may be observed simultaneously in a two channel instrument, or observations may be made of object and sky alternately.

Noise introduced by the detector itself can be of many kinds and will be discussed in the appropriate chapters. In photography and electronography the grain of the emulsion and its fog level are examples of detector noise. In electronic detectors such as photomultipliers and television tubes, noise manifests itself as random fluctuations in the output current, which could arise at any of the various stages of generation and amplification. The term 'dark current' is generally used to describe the current that is measured in such a detector in the absence of an input signal. The reduction of detector noise, for example by cooling photomultipliers, is a constant objective of the instrument designer.

For certain detectors, noise is introduced by a readout system. In the case of, for example, a SIT vidicon or a CCD, the readout is an integral part of the operation of the detector. In the case of a photographic plate being scanned by a microdensitometer, the noise characteristics depend on the machine in use.

For the remainder of this section, the term noise is used to include any of these various sources of signal degradation.

The concept of signal-to-noise ratio (S/N) is used to quantify the effects of noise. The input signal-to-noise ratio is the ratio of the input signal to the rms fluctuations in the input signal. In the example given above, the signal-to-noise is 3.2/1.74 = 1.84 for the 50 second observation. For any input photon flux following a Poisson distribution, the signal S is the mean k and the rms noise is $k^{1/2}$, giving $S/N = k^{1/2}$.

Each subsequent source of noise will reduce S/N, and a detector may be regarded as a collection of components each of which degrades the input signal-to-noise. The output signal-to-noise ratio is defined as the ratio of the input signal to the rms fluctuations in the output signal expressed in input signal units. Returning to the example, suppose a photomultiplier is set up to detect the incoming photons, and suppose its response is such that a steady flux of 10 photons per second would yield a d.c. signal of 3 pA. Assume that the output noise is measured at 0.7 pA. Then in input signal units, the output noise is $0.7 \times (10 \div 3) = 2.33$ photons per second. Since the input signal is 3.2 photons per second, the output signal-to-noise is 3.2/2.33 = 1.37. The observer is normally concerned with maximising the output signal-to-noise, even at the expense of worsening the noise.

The quantities S and N have been defined for signals which take discrete levels, but an analogous definition for continuous signals is quite straightforward. For a signal $y(t)$ which varies continuously with time t, $S = (1/T) \int y \, \mathrm{d}t$ and $N^2 = (1/T) \int (y-S)^2 \, \mathrm{d}t$, where the integrals are taken over the time interval T. If the signal fluctuates randomly such that the probability that the signal at any instant lies between y and $y + \mathrm{d}y$ is $p(y)\mathrm{d}y$, these definitions become $S = (1/T) \iint yp \, \mathrm{d}y \, \mathrm{d}t$ and $N^2 = (1/T) \iint (y-S)^2 p \, \mathrm{d}y \, \mathrm{d}t$ or simply $S = \int yp \, \mathrm{d}y$ and $N^2 = \int (y-S)^2 p \, \mathrm{d}y$, where the y integrals are taken over the whole range of possible y values. Here p is called the probability density, and $\int p \, \mathrm{d}y = 1$. The simple example of a d.c. signal with Gaussian fluctuations is described by the probability density

$$p(y) = \frac{1}{\sigma \sqrt{(2\pi)}} \exp\left(\frac{(y-y_0)^2}{2\sigma^2}\right)$$

It is easy to show that in this case the d.c. signal $S = y_0$ and the noise $N = \sigma$. It can be seen that the definitions of S and N are the first and second central moments of y respectively, that is, the mean and variance.

S and N describe two time-averaged properties of the signal, but it is often necessary to know what the time structure of the signal is, that is, whether certain frequencies are present in the fluctuations. To do this, the noise power spectrum $W(\omega)$ is used, defined as

$$W(\omega) = \left| \frac{1}{T} \int (y - S) \exp(-2\pi i \omega t) \, dt \right|^2,$$

which will be recognised as the square of the modulus of the Fourier transform of $(y - S)$. This function of frequency ω describes the relative strengths of the component frequencies of the noise fluctuations.

Although the above definitions have assumed a signal fluctuating in time, they all apply equally well to a signal varying in space, such as the density of a photograph as measured by a microdensitometer. The time variable t is simply replaced by the space coordinates, and the formulae are generalised to two dimensions in the obvious way.

One common problem is the detection of an object signal in the presence of a confusing background signal. Here we are interested in the differentiation of two signal levels, one with and one without the object, as illustrated in Figure 1.3. In this case we may take S as the difference in these levels and, on the assumption that the noise is the same at both levels, N as the noise of either of them. An S/N of three is commonly taken as reasonable confirmation of the presence of the object signal, and the expressions 'a three sigma detection' or 'detection at the three sigma level' are often used in this connection.

Figure 1.3. Detection of a signal in the presence of background noise.

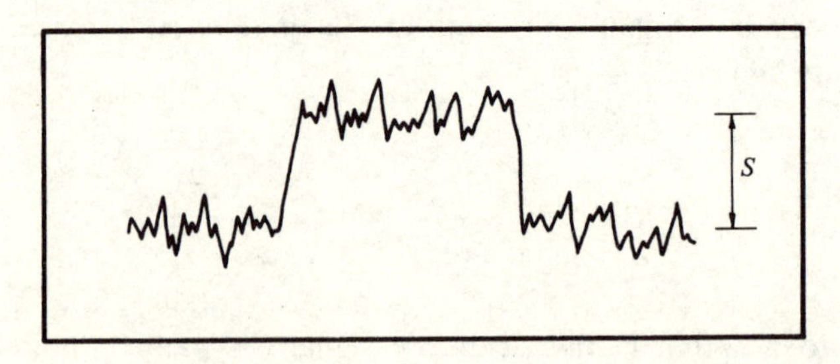

1.4 The assessment of detector performance

In this section, definitions will be given of certain concepts to be used in subsequent chapters to assess and compare the performance of different types of detector. They are common to all detectors, although some may be more appropriate to certain types than to others.

The first concept is the sensitivity of the detector. This may be simply defined as the ratio of the output signal to the input signal. Although useful enough in assessing the response of any one detector, the sensitivity gives no information about how efficient it is at recording photons. The quantum efficiency (QE) of a detector is defined as the number of photons recorded by the detector divided by the number of photons that would have been recorded by a perfect detector under the same conditions. This quantity is therefore a ratio and is usually expressed as a percentage. In a multi-stage detector, the quantum efficiency of each stage may be considered separately. The quantum efficiency is a useful general guide to the potential sensitivity of a detector, but is most often used in comparing the performances of detectors of the same type, for example the sensitivity of various photocathodes as a function of wavelength. The concept of quantum efficiency does not allow for any photons which may have been recorded and subsequently lost. It is therefore not very useful in dealing with the photographic emulsion or other detectors where it is difficult to quantify the numbers of photons detected.

The detective quantum efficiency (DQE) of a complete detector system takes into account any loss of detected photons and the quantum efficiency of each stage of the system. In a loss-free single stage system DQE and QE are the same, but DQE will be less than QE for most other systems. Neither DQE nor QE give any information about the quality of the detector output, since the precision with which a signal can be measured depends on its signal-to-noise characteristics. The detective quantum efficiency of a detector can be expressed in terms of the signal-to-noise characteristics of the input and output signals as the ratio

$$\mathrm{DQE} = (S/N_{\mathrm{out}})^2 \,/\, (S/N_{\mathrm{in}})^2$$

and is also usually expressed as a percentage. S/N_{out} will never exceed S/N_{in}, so the DQE represents the degradation of the signal by passage through the detector. Because of its general definition, DQE is a very useful parameter to compare the performances of different types of detector, for example a CCD and a photographic emulsion. Although one detector may be slower than another, that is, have a lower QE, it may nevertheless have better detectivity if the DQE is higher.

Spatial resolution is an important property of one- and two-dimensional detectors. One way to quantify this is to determine the output image density distribution resulting from an idealised input point image. This density distribution is called the point spread function, and typically has a profile rather like a Gaussian. The width of this profile, defined in some suitable manner, determines the resolution limit of the detector, since two output images will merge into each other if they are separated by less than the profile width. An analogous concept is the line spread function, the output image density distribution resulting from an idealised input line image. Each can be derived from the other, and the line spread function is generally easier to determine experimentally.

The spatial resolving power of a detector is often specified in terms of the number of line pairs per millimetre that it can resolve. This usually means that test patterns such as Figure 1.4 have been observed, consisting of arrays of alternating black and white lines or bars. The resulting image is examined to determine the highest spatial frequency that can be resolved. One line pair is one black and one white bar. This procedure is

Figure 1.4. A resolving power test pattern.

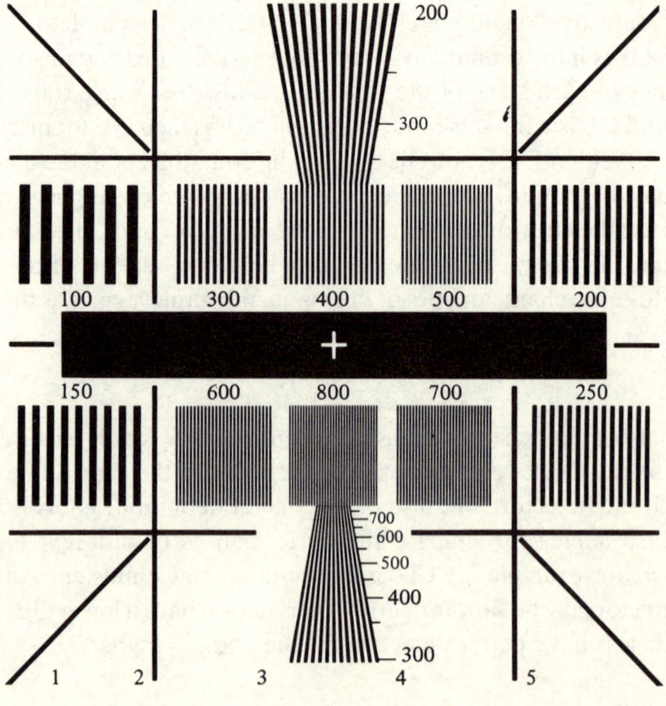

very straightforward but somewhat unsatisfactory in that the result can depend on the design of the test pattern, in particular on the contrast between the bars. A more objective measure of spatial resolution is the modulation transfer function, which is illustrated in Figure 1.5. Assume an image is input which takes the form of one-dimensional sinusoidal modulations, for example $E(x) = a + b \sin 2\pi\omega x$. The modulation of this image is defined as

$$M = (E_{max} - E_{min}) / (E_{max} + E_{min}),$$

which in the example is b/a. The output image will have a similar form but will be degraded by the detector to a lower modulation. The modulation transfer is defined as the ratio M_{out}/M_{in}, where M_{out} is evaluated in input signal units. This will be a function of the spatial frequency ω, falling from 100 per cent at low frequencies to zero at high frequencies, as illustrated in Figure 1.5. This function of ω is called the modulation transfer function (MTF).

Figure 1.5. An illustration of modulation transfer. A sinusoidal pattern such as that shown in (*a*) below is used as the input to a detector. A tracing of the detector output is shown in (*b*). The output pattern is of similar form to the input pattern, but is degraded to a lower modulation.

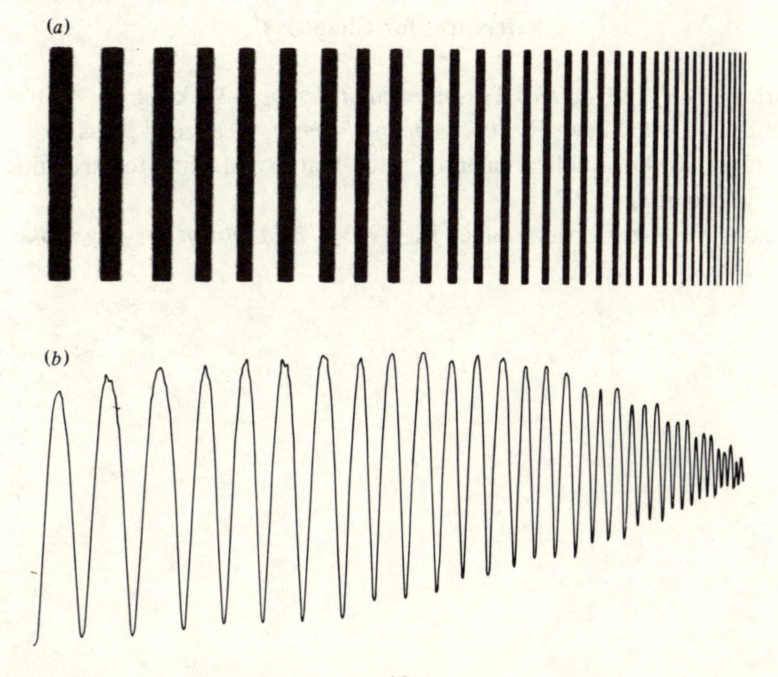

The resolving power is closely linked with the concept of a pixel. This can be defined as the size of the smallest element of a detector to which a definite signal level can be attributed. In the case of detector arrays such as CCDs, the meaning is obvious – each separate detecting cell is a pixel, and the pixel size is the spacing of the individual cells. It is not so easy to quantify the idea of a pixel for the photographic emulsion, but it could be interpreted as the reciprocal of the spatial resolving power. In other words, a resolution of 200 line pairs per millimetre implies a pixel size of 5 μm square. The information storage capacity of a pixel is the number of distinguishable signal values it can assume. These different values are called levels, and since each level is subject to uncertainty because of noise, the separation between them is set by some appropriate multiple of the noise. The number of different levels L can also be measured in bits of information, b, where $b = \log_2 L$. For example if there are 256 different available levels, each pixel can store 8 bits of information. If the pixel is exposed for a sufficiently long time, the output signal reaches the highest level L and can increase no further. This phenomenon is called saturation, and is common to all detectors, though it is easier to restrict the range of operation below saturation for some detectors than for others.

References for Chapter 1

Barlow, B.V., 1975. *The Astronomical Telescope*, Wykeham.

Dainty, J.C. & Shaw, R., 1974. *Image Science*, Academic Press.

Eastman Kodak publication no P-315. Plates and films for scientific photography, 1973.

Roach, F.E. & Gordon, Janet L., 1973. *The Light of the Night Sky*, Reidel.

2

The photographic emulsion as a detector

The photographic emulsion has been used as a detector in astronomical observations for over a century. The introduction of new emulsions specially designed for low light level conditions, and the successful application of pre-exposure hypersensitisation methods to these emulsions has led to the recent dramatic advances in astronomical photography. For some applications the photographic emulsion has been replaced by more sophisticated electronic detectors which are usually more efficient. But they are also more complex and expensive, and consequently may be beyond the reach of some smaller observatories. The light-sensitive photographic emulsion, coated onto a glass or film base, provides a relatively simple and inexpensive low light level detector. By careful selection of appropriate emulsions and colour filters from the considerable range available, radiation can be detected photographically from the atmospheric cutoff wavelength near 300 nm up to about 1.1 μm.

2.1 Advantages and disadvantages

There are a number of applications of the photographic detector in astronomy which cannot be undertaken by any known photoelectric device. The most important of these applications concerns imaging of large fields, such as those available in Schmidt telescopes and in Ritchey–Chretien systems. The UK Schmidt telescope at Siding Spring, Australia, and the Mt Palomar Schmidt in the United States each have a field of view of 6.5 degrees, and require a photographic plate 356 mm by 356 mm in order to record the full field. The Schmidt telescope of the European Southern Observatory in Chile has a 5.5 degrees field, and requires photographic plates 300 mm by 300 mm. The largest plate size in normal use is 500 mm by 500 mm, and is used at the Cassegrain focus of the Dupont telescope at Las Campanas in Chile. In comparison, electronic detectors rarely have dimensions exceeding 50 mm.

The Kodak IIIa-J emulsion is a typical fine-grain emulsion in regular use. It has been responsible for considerable advances in photographic recording since its introduction in the late 1960s, and is capable of a resolving power of 200 line-pairs per millimetre. It therefore has about 4×10^4 available pixels per square millimetre. Assuming each pixel can store six bits of information, that is, it is capable of registering 2^6 different density levels, then a 356 mm × 356 mm emulsion, as used in the UK Schmidt telescope, can store up to 3×10^{10} bits of information. Furthermore, defects in the emulsion occupy a very small proportion of the total area of the photograph. The large quantity of information recorded on photographic plates is stored permanently, and in a pictorial format which is readily accessible for visual study. For more quantitative analysis, however, the stored information is not immediately available for computer analysis. Powerful measuring machines or microdensitometers are required to read the information on the plate and convert it into digital form for computer processing, whereas most electronic detectors give a direct digital readout, sometimes during observation, so that the processed information is very quickly available to the astronomer for analysis and display.

The dimensional stability of a photographic plate, combined with the large field and permanence of the record, allow very accurate position measurements to be made of a large number of objects on one plate. This is particularly valuable in astrometric studies, or in photographic spectroscopy if an electronic detector is not available. Some of the new film bases, especially the thick Estar base, approach the dimensional stability of glass to within a factor of two. Since the expansion coefficient of glass is about a third that of hardened steel, the thermal dimensional stability of a photograph on a glass or film base may well be better than that of the machine used to measure the photograph. It has been suggested that the photographic emulsion may distort during processing, causing small shifts in image positions, but measurements to the precision required for astrometry have been made on large Schmidt photographic plates which were carefully and correctly processed, and no quantitative evidence for such emulsion shifts has been found.

The photographic emulsion suffers from several disadvantages compared to electronic detectors. It is very inefficient. Even the best emulsions have a peak detective quantum efficiency (DQE) of only 4 per cent as compared with the charge-coupled device (CCD) which may have a DQE of 60 per cent. Consequently photographic observations require considerably longer exposure times than those required for electronic systems. The response of the photographic emulsion to light is not linear, although this can be overcome to a considerable extent by careful calibration of each plate during exposure. The uniformity of response over the large area of a photographic plate is often criticised, but in fact the intrinsic uniformity of the emulsion is generally far greater than for electronic systems and may be as good as 2 per cent on scales of 10 cm. The problem is that it is very difficult, and often impossible, to correct for photographic non-uniformity by flat-field exposure methods, whereas these techniques are relatively simple to apply to electronic detectors. Non-uniformity of the photographic response over a single plate, or from one plate to another of the same type, may be introduced by the processing procedures required to convert the exposed emulsion to a permanent and visible record. Other photographic effects, such as the Eberhard effect, adjacency and edge effects, may also be introduced during processing. Most of these difficulties can be reduced or eliminated by paying careful attention to every step of the processing procedures. The processing of photographic plates and films is always a potentially messy procedure, since every stage of the procedure involves a liquid.

The photographic emulsion is most effective and useful when applied to wide-field work, such as sky surveys, and in other instances when good resolution is required over a relatively large area, as for instance in stellar photometry of large numbers of stars, surface photometry of galaxies and extended nebulae, astrometry, and provision of large samples of data from which statistical studies can be made. New techniques made possible by the combination of large photographic plates, high speed measurement techniques and powerful computing facilities include the automatic separation of galaxy images and star images. There is no electronic device which can combine the benefits of high resolution and large format to the extent available using the photographic plate.

2.2 The characteristic curve

The degree of blackening of a photographic emulsion due to exposure to light and subsequent chemical processing is directly related to the intensity of the exposing light source. The characteristic curve of the emulsion–developer combination, sometimes referred to as the H & D curve after Hurter and Driffield who first used it, is the graph of the densities on the processed emulsion plotted against the logarithm of the exposures which caused them. Densities which are due to sources other than the signal being recorded are sometimes referred to as fog.

The characteristic curve is typically of the form shown in Figure 2.1, where D is diffuse density and E is exposure. There is an unavoidable minimum density D_{min} or gross fog which will occur even without exposure to light. This consists of base fog, due to the imperfect transmission of the glass or film upon which the emulsion is coated, plus the chemical fog resulting from those emulsion grains which have been developed even though they have not received any light. The chemical fog may depend on both the emulsion and the developer used. The region A to B, referred to as the toe of the H & D curve, corresponds to the minimum exposure levels which will give a density above gross fog. The non-linearity of the curve in this region of low exposure gives rise to problems in the calibration of photographic materials. The region B to C is the usual range of operation of an emulsion, where the relationship between density and log E is most nearly linear. The latitude and dynamic range of an emulsion are the ranges of exposure and density respectively which correspond to the linear part of the characteristic curve. The region C to D, the shoulder of the curve, is again non-linear, and corresponds to progressive saturation of the emulsion.

The slope $dD/d \log E$ of the straight-line portion of the characteristic curve between points B and C is the contrast or gamma (γ) of the emulsion. At any point P on the characteristic curve an instantaneous or local value of gamma may be obtained from the slope of the tangent to the

Figure 2.1. The form of the characteristic curve for a photographic emulsion. The contrast or gamma of the emulsion is the slope of the straight-line portion of the curve. The local gamma at any point P is obtained from the slope of the tangent to the curve at that point.

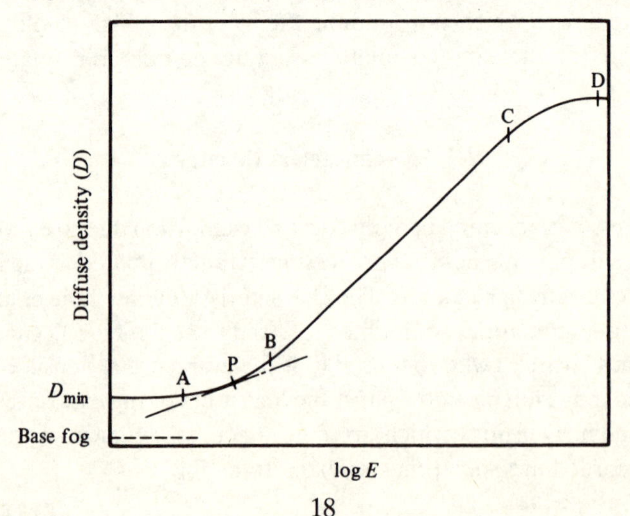

characteristic curve at that point. There is another parameter, contrast index (CI), which should not be confused with gamma. The CI describes a generalised type of contrast, and may include parts of the toe and shoulder of the characteristic curve. It is meant for use in pictorial photography, and is not applicable to astronomical photography.

Photographic density is measured using a densitometer, and may be defined as follows. If a light flux F_{in} is incident on the specimen in the densitometer and the flux transmitted through the specimen is F_{out}, then the transmittance of the specimen is $T = F_{out}/F_{in}$, the opacity O is $1/T$, and the transmission density is defined as $D = \log_{10}(O)$. If some of the transmitted flux is lost and is not recorded by the densitometer, then the measure will be of specular density. A measurement of density in which all the transmitted flux is gathered and recorded by the densitometer is called diffuse density, and there is a standard specification of diffuse density (American National Standards Institute Standard PH2.19, 1959 and 1975). The difference between diffuse and specular density is illustrated in Figure 2.2. The above definition of diffuse density implies that the input light is collimated and that all the output flux is collected. But since optical paths are reversible, the opposite arrangement is also

Figure 2.2. The difference between diffuse and specular density. If the beam incident on the sample is collimated, and all the transmitted flux is collected, diffuse density is measured. If some of the transmitted flux is not collected, specular density is measured. The reverse is also true. If the transmitted flux is collected into a collimated beam, diffuse density will be measured if the incident light is perfectly diffuse, and specular density will be measured if the incident light is not perfectly diffuse.

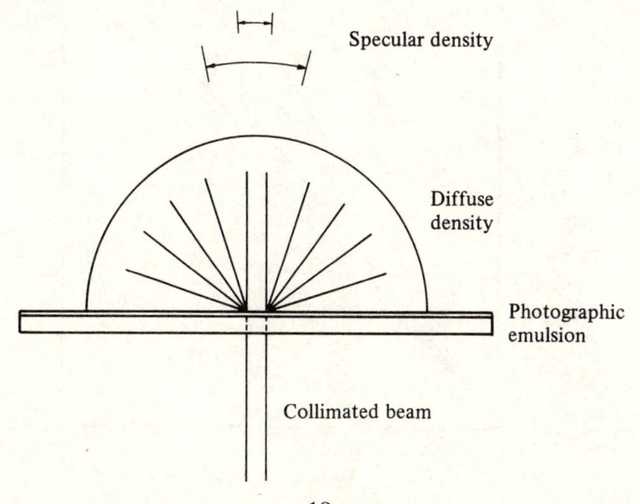

true, and is more usual in practice. It is useful to remember that 0.3 diffuse density corresponds to 50 per cent transmission, and 1.0 diffuse density corresponds to 10 per cent transmission. The zero point of the density scale is defined by having no specimen in the densitometer, corresponding to 100 per cent transmission.

It is not difficult to decide whether a densitometer will give specular or diffuse values of density by inspection of the optical system. For example PDS (Photometric Data Systems) and Joyce Loebl microdensitometers both lose some flux and will therefore give specular measures, whereas the Macbeth and Densichron densitometers collect all the transmitted flux and therefore give diffuse density measurements. The proportion of flux lost will differ from one type of densitometer to another, and it is therefore not practicable to define a standard of specular density. Specular density measures may be converted to diffuse measures via a reference sample with the same grain characteristics as the emulsion being measured, and with several different densities on it, which has previously been measured by a diffuse densitometer. As shown in Figure 2.3, the relationship is approximately linear. The ratio Q of specular measure to diffuse measure is called the Callier Q factor, and is the slope of the calibration plot. Q increases with the specularity of the densitometer, and

Figure 2.3. The Callier Q factor indicates the degree of specularity of density measurements. For a given emulsion sample, Q is the ratio of the density measurements obtained from a specular densitometer to those obtained for the same points from a diffuse densitometer.

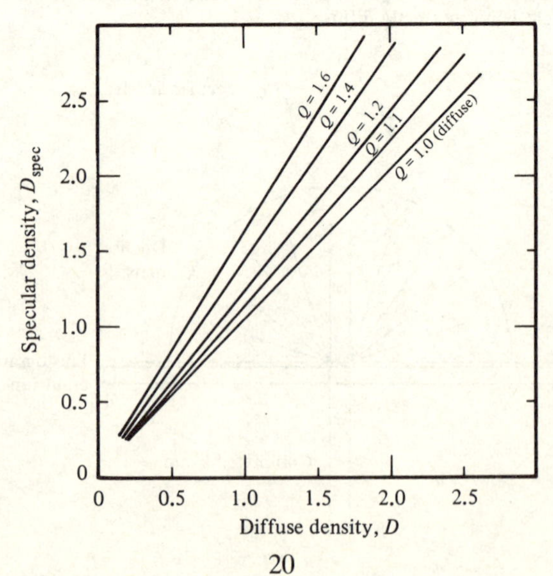

also with granularity of the emulsion, and must therefore be evaluated for each densitometer and emulsion type being used. For measures on a particular specular densitometer, the Q factor will be greatest for measurements of coarse-grained emulsions and least for measurements of fine-grain emulsions, as shown in Figure 2.4. In many laboratories and observatories it has been customary to make all the density measures on a scale of specular density. Within the particular laboratory this is valid if the system is internally consistent. However, such a laboratory cannot compare its findings directly with other laboratories, or with the manufacturer's data, unless the laboratory measures are expressed in terms of diffuse density. Density values in this book and most other recent publications concerning astronomical photography are expressed in units of diffuse density unless otherwise stated.

The exposure value E takes into account both the time t during which the emulsion was exposed to the light source, and the illuminance I received at the emulsion from the source. E is defined as $E = I \times t$, and may be given in photometric units (lux seconds), or in units of incident energy per unit area (joules per square metre), or as photons per unit area. It is usual to plot log E rather than E, although there are some significant papers in the literature which use E rather than log E. (Throughout this book log means logarithm to the base 10.) The data in most commercial publications are given in absolute units of exposure, but there are as yet very few observatories which have the capability to

Figure 2.4. The Callier Q factor is higher for coarse-grained emulsions than for fine-grained emulsions. Consequently specular density measures lead to higher apparent values of speed, gamma and chemical fog density than those obtained from diffuse density measures of the same sample.

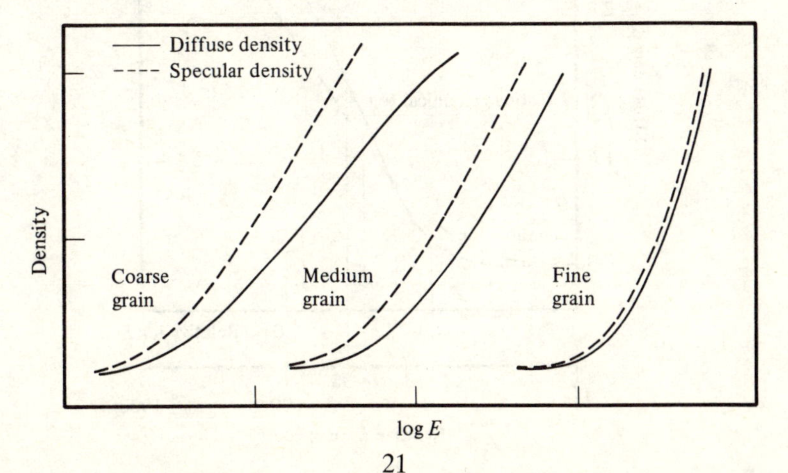

measure absolute values. Consequently H & D curves are plotted against relative scales of log exposure. Since density is also a logarithmic function, the H & D curve is a log–log plot, and gamma is a ratio of two logarithmic quantities.

The speed of an emulsion for astronomical purposes may be usefully defined as the exposure E required to attain a specified density above chemical fog with the chosen combination of telescope, emulsion, colour filter and processing conditions. When defined in this way speed is a direct function of the required exposure time. By contrast the more familiar ASA and DIN definitions of photographic speed are inverse functions of exposure time, and in any case are not applicable to the long exposure times used in astronomical photography. The density level at which the speed is to be measured will depend upon the intended application of the material. To optimise the output signal-to-noise of plates taken with fast Schmidt cameras, where the night sky brightness contributes significantly to the background density, the point at which the speed is measured – that is, the speed point – may be at diffuse density 1.0 above chemical fog. In Figure 2.5, both emulsions have a chemical fog level of 0.3 diffuse density units. The speed point is therefore at density 1.3 diffuse. At this point the measured speed of emulsion A is 35, and of B

Figure 2.5. The use of a speed point as a measure of emulsion response.

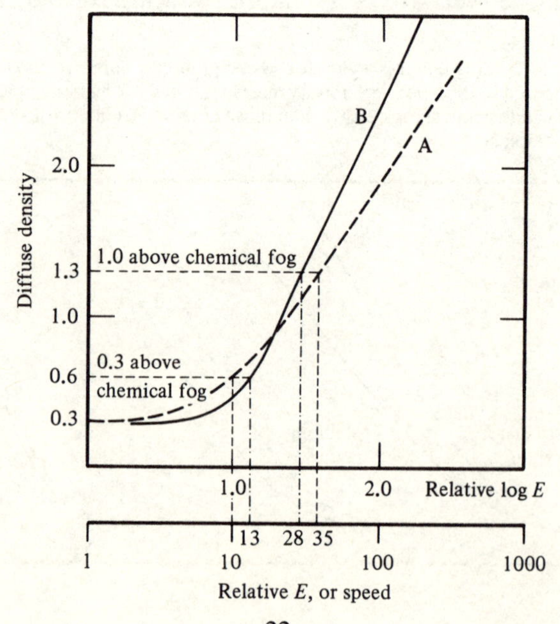

is 28; emulsion B will reach density 1.0 above the chemical fog with 20 per cent less exposure than emulsion A. For photographic recording of the output from a spectrograph, speed is more commonly assessed at density 0.3 above chemical fog. Referring again to Figure 2.5, the speed point in this case will be at density 0.6 where the measured speed of emulsion A is 10 and of B is 13. In these conditions emulsion A will reach the required density with less exposure than emulsion B. However, the speed of the available emulsions is not necessarily the most important parameter in the selection of an emulsion for a particular application. Selection will depend on several factors, amongst which the signal-to-noise behaviour of the emulsion is one of the most important.

The characteristic curve is a powerful and versatile tool with which to investigate the performance of emulsions subjected to different development systems, or to different methods and degrees of hypersensitisation, as well as providing speed measures and estimates of exposure time. It also provides a foundation for the derivation of other important information such as the signal-to-noise behaviour of the emulsion.

2.3 Factors influencing the characteristic curve

The behaviour of an exposed and processed emulsion is very sensitive to a number of factors. The characteristic curve reflects the combined results of all of these factors, some of which are under the direct control of the user and others of which are a consequence of the basic physics and chemistry of 'the emulsion and the manufacturing processes. In this second category, the user's control is limited to the selection of the emulsion to be used from those types made available by the manufacturer.

The manufacturing factors which have most influence on the emulsion characteristics are the chemical composition and dye-sensitisation of the emulsion. Its natural sensitivity to ultraviolet and blue light may be extended towards longer wavelengths by the inclusion of sensitising dyes. The mean grain size and the range of grain sizes created during production of the emulsion will determine its granularity and will strongly influence its intrinsic speed and contrast class. In general fine-grain emulsions have high contrast and resolution, but are relatively slow.

The gamma and the intrinsic sensitivity of an emulsion also depend on the wavelength of the exposing light-source. The sensitivity of the IIIa-F

Figure 2.6. The variation of emulsion sensitivity with the wavelength of the exposing light, for a sample of type IIIa-F emulsion. (a) The sample is exposed in a calibration spectrograph to give a set of spectra corresponding to different light intensities. (b) Microdensitometer tracings of the spectra on the sample show the variations of density with the wavelength of the exposing light. (c) Characteristic curves may be constructed from these tracings, to obtain the variations in gamma with wavelength.

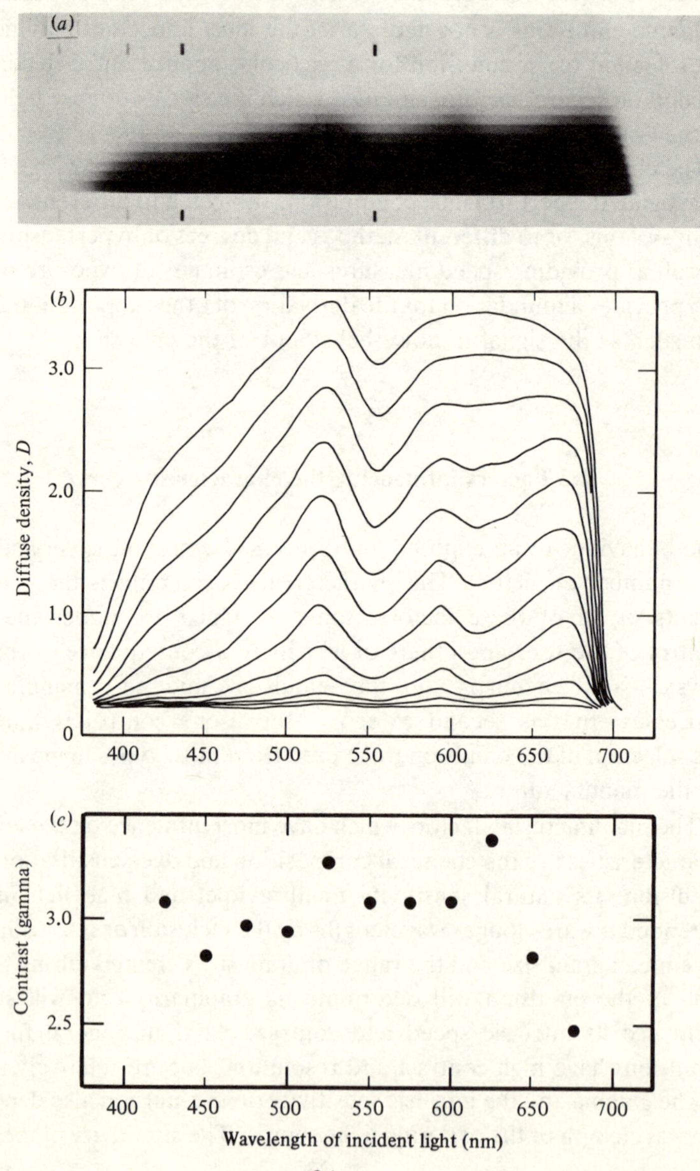

(dye-sensitised) emulsion is shown in Figure 2.6 as a function of wavelength, and it can be seen that the sensitivity to green light is less than to blue or red light. A sample of the emulsion of interest is exposed in a calibration spectrograph (see Section 2.6) to a set of spectra of different intensities (Figure 2.6a) and the response of the emulsion can then be measured in terms of the output density as a function of wavelength and intensity. The mean intensity interval between spectra in Figure 2.6 is 0.140 log I units. Characteristic curves of the emulsion response at different wavelengths may be derived from these spectra, and clearly they will depend on wavelength. In particular, the gamma changes with wavelength, as indicated in Figure 2.6c. Thus the response of an emulsion exposed to a narrow range of wavelengths, as for instance through a narrow-band interference filter, may not be the same as the response to an exposure to a broad range of wavelengths with the same mean wavelength. For this reason an emulsion under test or one used for calibration purposes must always be exposed to light of the correct wavelength and wavelength range. It is particularly important to take into account these wavelength variations of emulsion response in critical applications such as photographic spectroscopy and narrow-band photometry.

Immediately after an emulsion is coated onto its film or glass base it begins to age, and will continue to do so until it has been processed. The rate at which an emulsion ages depends significantly on the conditions in which it is stored. The symptoms of ageing are gradual loss of speed and increase of chemical fog, and there is no really effective way to overcome effects once they have appeared. Even unexposed, an emulsion has enough developable grains to produce some fog density, which should not exceed about 0.1 on a fresh emulsion, but the number of these grains and the chemical fog level increase slowly with age. The fog may also increase if the emulsion is exposed to chemical contaminants or allowed to become too warm. Exposure to oxygen and water vapour, together or separately, decreases the sensitivity and therefore the speed of an emulsion, and causes an accompanying increase in the chemical fog level. Known potential chemical contaminants include inadequately cured cardboard, especially if it is damp, some foam plastics, and the vapours given off by some types of paint, especially those which are oil based. These effects can occur during shipment from the manufacturer, as well as in storage of the plates at the observatory. In order to slow down the rate at which an emulsion shows these signs of ageing, it is recommended that spectroscopic emulsions should be stored in a dry, oxygen-free environment, in suitable chemically inert containers, inside a cold room at approximately 5 °C, or a deep freeze at approximately −20 °C.

Whilst hypersensitisation usually improves the response of an emulsion, it is often at the cost of an increase in chemical fog, and a small reduction of gamma and signal-to-noise. It is not yet certain if and to what extent hypersensitisation may alter the wavelength dependence of the emulsion sensitivity. The IV-N emulsion hypersensitised in silver nitrate shows a larger gain in sensitivity in the red, dye-sensitised, region than in the blue region, but the degree of this differential hypersensitising is rather unpredictable. As a result an exposure on a hypersensitised emulsion should not be calibrated from an exposure made on unhypersensitised emulsion, even of the same batch, especially in critical applications which require quantitative results of high accuracy.

The ambient atmosphere and temperature during an exposure can also affect the characteristic curve. Some emulsions may show an apparent increase in speed at reduced temperatures, but others show the opposite effect. Although it is clearly impossible to expose all plates at the same temperature in the telescope, it is important that exposures to be compared quantitatively – especially a telescope plate and its calibration plate – should be undertaken as nearly as possible at the same temperature. The presence of oxygen or water vapour during exposure is known to reduce the effective sensitivity and contrast of an emulsion, often with a corresponding increase in the chemical fog level. Some observatories have dealt with this problem by converting the plateholder into a gas-tight cassette, with the filter sealed into position, which can be filled with dry nitrogen during the exposure. As a result the emulsions can sometimes operate with effective speeds twice as fast as obtained in normal plateholders, allowing much more efficient use of the telescope. In humid conditions the exposed emulsion should be stored in a dry nitrogen atmosphere until it is processed, in order to avoid latent image decay and loss of sensitivity caused by the ambient atmospheric oxygen and water vapour.

The characteristic curve shows the cumulative result of all the parameters which affect it, and the disentanglement of the various factors and their effects is not always straightforward. It is remarkably easy to attribute subtle changes in the emulsion performance and the characteristic curve to storage or hypersensitisation instead of the processing, or vice versa.

Of all the factors which can influence the characteristic curve, development is the most important. The development stage of photographic processing is the chemical amplification of the weak recorded input signal. It is the direct analogue of the amplification stage in an electronic system. As with all amplifiers, small changes in the design or operation of the amplifier can result in significant changes in the output signal. The rate at which the latent image on an exposed silver halide grain is amplified, or converted into image silver, depends on the activity level of the developing solution and the rate at which fresh developing agent is made available to each grain. The level of agitation of the developer solution controls the rate at which partly-used developer is removed from contact with the emulsion and is replaced by fresher, more active solution. An energetic developer such as D19 requires only 4 or 5 minutes to develop emulsion type IIa-O to its optimum level, but D76, a less energetic developer, requires 15 minutes to obtain a similar result. Dilution of the developer also reduces the rate of development. Kodak Technical Pan Film 2415 (previously called SO-115) is particularly sensitive to variations in developer activity levels, and to solutions of different strengths, as shown in Figure 2.7. The variation of contrast with developer strength is particularly marked. All photographic emulsions are subject to these

Figure 2.7 The dependence of the characteristic curves obtained for Kodak Technical Pan Film 2415 on the type and dilution of the developer, and on the development time. HC110-F is a weaker solution than HC110-D.

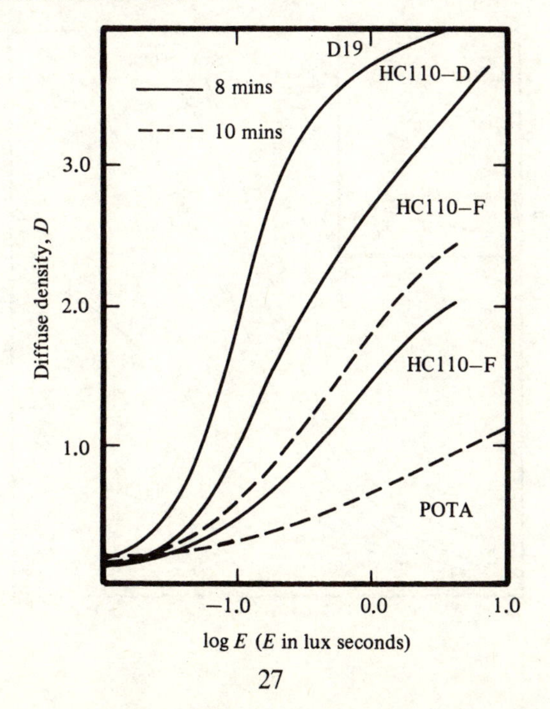

effects, and although they are not so dramatic on other emulsions, they are still very significant. The wide range of characteristic curves which can be obtained from the 2415 emulsion is one of the primary reasons for its importance in the photographic laboratory.

The rate of reaction of any chemical process depends on the temperature at which the reaction takes place. Photographic processes are no exception to this rule. It is therefore standard practice to develop exposed emulsions at 20 °C, preferably in an air conditioned darkroom to ensure stable conditions throughout the processing sequence. For critical work, where plates or films must be processed identically, the temperature of the developer is maintained to ± 0.1 °C. Development is the only one of the photographic chemical processes which is not taken to completion, since this would result in a totally black emulsion. It is therefore necessary to time the process very carefully, and to terminate the development effectively at the end of the required time. Development is only possible in an alkaline solution, and therefore development can be terminated by immersing the emulsion in a stop bath of one per cent acetic acid solution.

Figure 2.8. The dependence of the characteristic curve on development time for a type 103a emulsion. The effect on the gamma and on the chemical fog density level is shown in the inset.

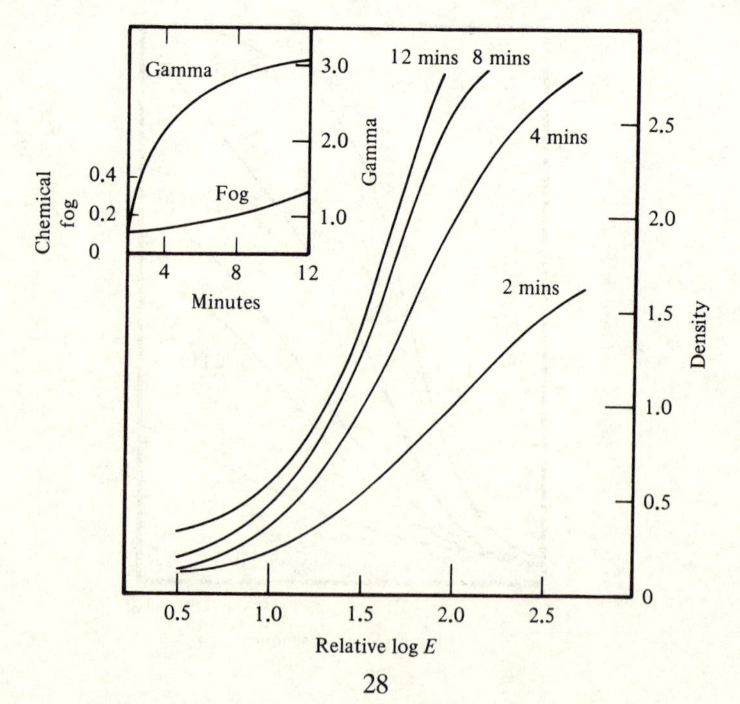

The dependence of the characteristic curve on development time for a typical type 103a emulsion developed at 20 °C is illustrated in Figure 2.8. Increased development time or temperature normally results in increases in gamma and in chemical fog level, although this is not so marked in the case of the IIIa-J emulsion. It does not automatically follow that the speed, measured at a given density above chemical fog, will increase with development time, since the speed point will be forced to higher total densities as the chemical fog level increases. Indeed over-development can result in losses of output signal-to-noise and DQE.

The effects on the characteristic curve of low intensity reciprocity failure and of hypersensitisation will be discussed more fully in Sections 2.4 and 3.2 respectively.

2.4 Low intensity reciprocity failure

In 1862 Bunsen and Roscoe proposed a general law for photochemical reactions, which stated that the quantity of a photoproduct in a reaction depends only on the total energy employed. In the photographic case, this would mean that the total amount of photolytic silver produced by the incident light would depend only on the total exposure $E = It$. A common interpretation of the Bunsen–Roscoe law is the expectation that the density of the processed image would also depend only on the product It, and not on the individual values of I and t. This is known as the reciprocity law.

In situations where the exposure time is less than about 1 second it is true that an exposure at double the intensity for half the time would have the same photographic effect. There are however several additional processes involved between the production of photolytic silver and the final density of the processed material, so that the total quantity of photolytic silver produced is not necessarily related directly to the density of the final image. In any case, in astronomical photography the intensity of the incident light is usually very low, and the exposure times are very long, typically from 5 minutes to 5 hours. In these circumstances the reciprocity law fails and the individual values of I and t become important. If the reciprocal relationship of illuminance I and exposure time t is valid, then a plot of density D against total exposure E for different values of I and t, but constant It, should result in the straight line shown in Figure 2.9, where exposure time increases to the left, illuminance increases to the right, and the product $E = It$ is constant at all points on this axis. In practice the result is not the expected horizontal line, but a curve, such that at extreme values of I and t the resultant density is lower than expected, and the emulsion is operating less efficiently under these conditions. In photography in general, and in astronomical photography in particular, this failure of the reciprocal relationship between I and t is more the rule than the exception. Solar astronomers may have to deal with failure at high intensities, but the stellar astronomer has to accept low intensity reciprocity failure as a significant fact of life. There are techniques which can be used in the manufacture of the emulsion, and others (described in Section 3.2 on hypersensitisation) which may be used before exposure of the emulsion at the telescope, which can help to ameliorate the problem.

In order to include all three parameters E, I and t, the usual presentation of the effect of reciprocity failure is to plot log E as a function of log I at a specified exposed density, as for example in Figure 2.10. Lines of

Figure 2.9. At extreme values of intensity and exposure time the image density is lower than expected, due to the failure of the reciprocity law. The broken line indicates the expected density level, and the solid curve indicates the actual image density obtained.

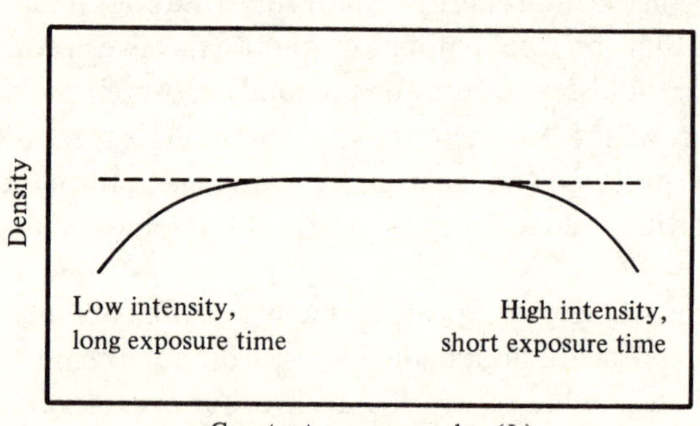

Figure 2.10. The reciprocity failure curves for Kodak spectroscopic emulsions types 103-F and 103a-F. These emulsions are essentially identical except that the 103a-F has been treated in manufacture to reduce low intensity reciprocity failure and the 103-F is not so treated.

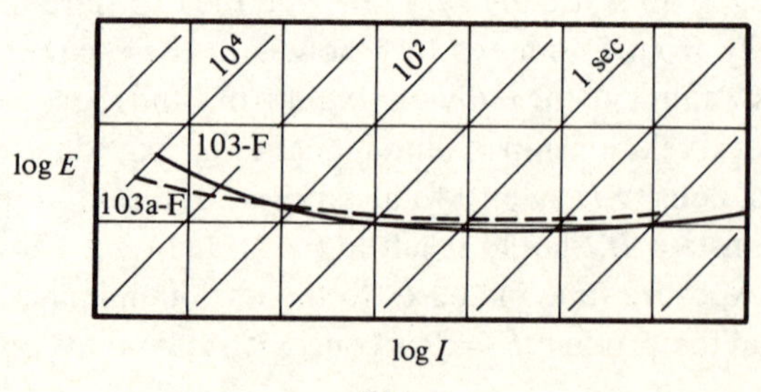

constant exposure time correspond to parallel straight lines at 45° to the log E and log I axes. Occasionally a similar plot of log E versus log t is used, in which case the diagonals represent lines of constant illuminance. In each case the region of optimum efficiency of the emulsion is that part of the curve corresponding to lower values of It. The portion of the curve (if any) parallel to the log I (or log t) axis corresponds to the limited range of values of I and t for which the reciprocity law is valid. Different types of emulsion may have very different reciprocity failure characteristics, and consequently it is sometimes possible to select an emulsion for which reciprocity failure is minimal at the intensity and exposure levels required in a particular application. For example, Kodak spectroscopic emulsions including 'a' in their designation have been specially treated during manufacture to reduce reciprocity failure for long exposures. Therefore, referring to Figure 2.10, type 103a-F (treated) is more suitable than type 103-F (not treated) for astronomical applications.

Reciprocity failure depends on the temperature of the emulsion during exposure. As the temperature is reduced below room temperature, reciprocity failure is also reduced, but below a certain temperature the emulsion sensitivity also falls. When the temperature is low enough to eliminate reciprocity failure, the response of the emulsion is very much slower than at room temperature. Reciprocity failure also varies irregularly with the wavelength of the incident light. Colour film generally consists of several layers of colour filters and emulsions, but

Figure 2.11. In a colour film the reciprocity behaviour of each emulsion layer is different, and consequently the colour balance of the film will vary with exposure time. In this example, as the exposure time increases, the blue-sensitive layer loses its sensitivity faster than the other layers do, and the resulting picture will be too yellow.

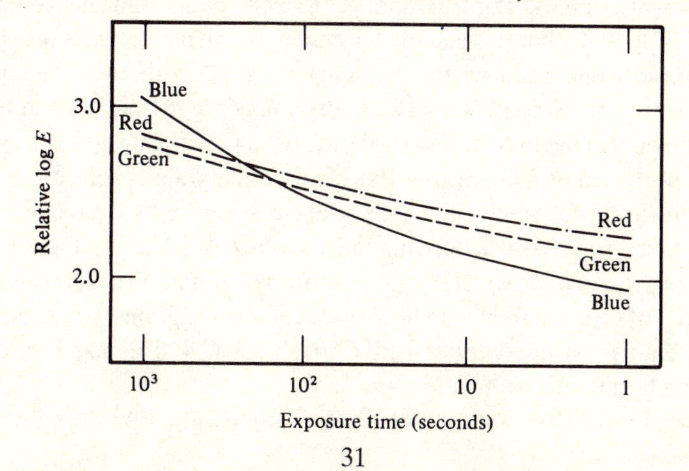

each layer is sensitive to a different colour and exhibits very different reciprocity law behaviour, as shown in Figure 2.11. Consequently if colour film is used outside the range of reciprocity law validity, the colour balance will be grossly distorted.

Test or calibration exposures must be made under the same conditions of reciprocity failure as the photograph to which they will be compared. This means that both exposures should be for the same exposure time. Only if the reciprocity law is known to be valid in the range of time and intensity to be used – that is, the reciprocity curve is parallel to the log I axis throughout this range – is it safe to use different exposure times and intensities for the test and comparison exposures.

2.5 Latent image formation

The principal constituents of the photographic emulsion are crystals of light-sensitive silver halide (grains) suspended in gelatin, together with chemical or colour sensitisers to give the emulsion a particular response characteristic, and a preservative or restrainer to ensure a commercially realistic shelf-life of the product by restraining the growth of chemical fog. The supporting medium for the light-sensitive grains is required to keep individual grains separated and well dispersed. It must be stable, so that both the unprocessed and the processed emulsion will be reasonably permanent. It must also be sufficiently porous to allow developing chemicals to reach the grains, and it must not interfere with the photographic process. For most purposes gelatin is still the most suitable medium, although its photographic properties may depend significantly on its previous history, so that photographic gelatin is selected from a limited number of sources. Pure gelatin swells very easily and may begin to dissolve at 40 °C, but it is usually hardened, which increases its resistance to abrasion and reduces the tendency to swell without reducing the porosity to processing chemicals. It also raises the temperature at which the gelatin will begin to soften to about 70 °C. Gelatin absorbs ultraviolet light shortward of 200 nm, but the silver halide grain is still sensitive to very much shorter wavelengths. Special emulsions can be made based on those first designed by Schumann in 1895 which contain very little gelatin and are sensitive to wavelengths as short as 4.4 nm, although the small quantity of gelatin offers very little protection to the grains. Consequently these emulsions are very sensitive to abrasion and damage, and are extremely difficult to handle.

The light-sensitive component of the photographic emulsion consists of fine crystals or grains of silver halide, which form 30 to 40 per cent by weight or 10 per cent by volume of an average emulsion. The most usual halides are silver bromide (AgBr) or silver chloride (AgCl). Both of these form a simple face-centred cubic crystal lattice like sodium chloride, with positive silver ions and negative halide ions arranged symmetrically as shown in Figure 2.12 so that each ion is surrounded in space by six ions of the other type, and no one halide ion is uniquely related to any one silver ion. The distance between an ion and its nearest neighbour is typically 2.8 $\times 10^{-4}$ μm, and a grain or crystal in a typical emulsion may be from 0.1 to 1.4 μm across. An emulsion in which the grains are of relatively uniform size is described as monodisperse. Both the mean grain size and the range of grain sizes are important factors influencing the intrinsic sensitivity or speed of the emulsion and its contrast and resolution. In general larger grains collect photons more easily than smaller grains and emulsions with a wide range of grain sizes have a greater range of sensitivity (that is, a greater latitude) than do monodisperse emulsions, although in many cases these apparently more sensitive emulsions also have lower contrast and poorer resolution. In considering the microscopic behaviour of a photographic emulsion and in particular those processes involving the grains, it is important to remember that the grains are separated from each other by gelatin and consequently every grain acts completely independently of every other grain. Each is a single unit upon which latent images may form, and each is developed independently of the others. The final image is thus composed of a collection of many individual grains of silver.

There are a number of distinct stages between the arrival of photons at the emulsion and the completion of the final permanent record. The release of a photoelectron by absorption of a photon at the grain is the

Figure 2.12. The face-centred cube structure of a silver halide crystal.

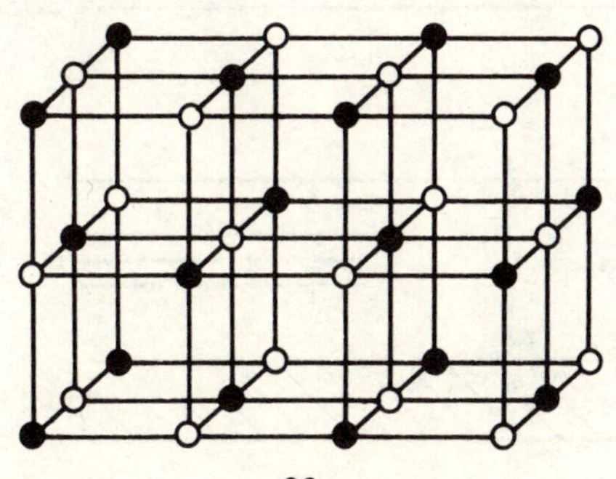

first of a group of processes which build up microscopic particles of metallic silver on the grains, known as the latent image. The latent image is chemically amplified, or developed, until it is large enough to be visible. Further chemical processing – fixation and washing – is required to remove any undeveloped light-sensitive material and other byproducts, leaving a permanent final record. The rest of this section will deal with the possible mechanisms by which the latent image is produced, in order to obtain some insight into the behaviour of the photographic emulsion as a light detector.

The photographic crystal is essentially a solid state detector, the properties of which will be more fully discussed in Chapter 5. A crystal of silver halide behaves like an insulator, in that the conduction band is normally completely empty, and the energy-level difference between the highest filled band and the conduction band is about 2.5 eV. Figure 2.13 shows an energy-level diagram for the silver halide crystal. Consequently thermal fluctuations alone are not normally enough to raise an electron from the uppermost filled band into the conduction band. However, excitation of a bromide ion by a light quantum of the necessary energy can drive an electron from the uppermost filled (valence) band into the empty conduction band, whence it is able to travel through the crystal as easily as through a metal. In addition to the photoelectron, there are two other mobile populations within the crystal lattice. The region vacated by the photoelectron has an excess positive charge, and is known as a positive hole. If such a positive hole can capture an electron from a neighbouring bromide ion, the site of the positive hole migrates within the crystal. The displacement of a silver ion from its place in the crystal lattice requires only 1.27 eV, and this is low enough to permit the production of a small

Figure 2.13. The energy-level diagram of a silver halide crystal.

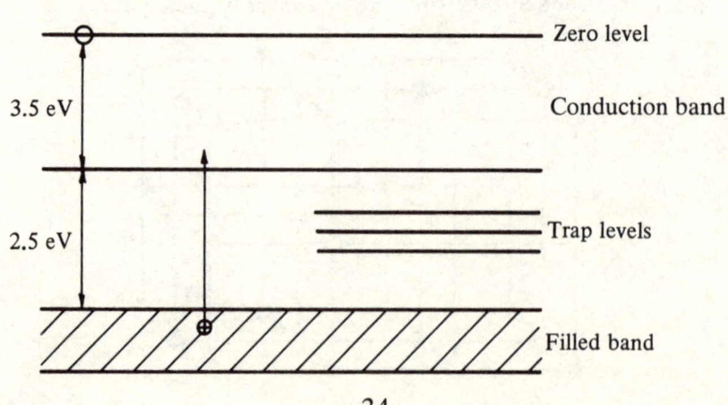

number of interstitial silver ions at room temperature which are also moderately free to migrate through the lattice structure.

It has been shown experimentally that the sensitivity of perfectly formed silver halide crystals is less than that of a crystal containing defects, implying strongly that the defects contribute significantly to the sensitivity of the crystal by assisting in the formation of the latent images. The defects occurring in a crystal may be due to displacement of individual ions (point defects) or to displacements or distortions of planes of ions (dislocations). Several experiments and demonstrations have shown that latent image silver tends to form preferentially at defect sites and particularly at dislocations within the crystal. It seems likely that these dislocations act as electron traps.

A photographic emulsion normally contains a chemical sensitiser, and perhaps also a sensitising dye. The actual mechanisms by which these sensitising agents operate on the grain are uncertain, but there is evidence to suggest strongly that 'sensitivity specks', sometimes of silver sulphide, are formed at various places in the crystal and that electrons and silver ions are preferentially attracted to these specks.

Although theories differ as to the finer details of the processes leading to the formation of the latent image, it is generally agreed that after the capture of a mobile photoelectron or silver ion at a sensitivity speck or at a defect site, electrons and silver ions are captured alternately, thus building up a group of neutral silver atoms to form the latent image. Not all the photoelectrons reach a site of latent image formation. Many will be caught in non-productive traps such as positive holes. It is also possible that halogen atoms may recombine with photoelectrons, with a corresponding release of energy, or even that the halogen atom may act as an oxidising agent and remove a photoelectron from the silver speck. Experiments indicate that a latent image must contain an aggregate of at least three silver atoms if it is to be stable against thermal decay. The relative mobilities of the three main components are significant. The most mobile are the photoelectrons, then the positive holes, and the least mobile are the interstitial silver ions. These differences contribute to the probabilistic nature of latent image formation. A certain amount of control over mobilities and other variables is possible during the manufacture of emulsions for various applications.

It can now be seen that the failure of the reciprocity law at low light levels is a direct consequence of the slow rate of arrival of photons, since the rate of production of photoelectrons, and consequently the rate of growth of the latent image, may not be fast enough to counteract processes by which the latent image may decay or be eroded. If the interval between the arrival of successive photoelectrons at a latent image site increases, so does the probability that the latent image may decay thermally. The problem of low intensity reciprocity failure will therefore persist unless it can be partially alleviated by increasing the intensity of the incident light (for instance, by using a faster camera system or telescope focal position) or by improving the stability of the sublatent image against thermal decay.

At the other end of the intensity scale, high intensity reciprocity failure occurs despite a plentiful supply of photoelectrons. In this case the mobility of the ions is such that the photoelectrons arrive at an image centre faster than the silver ions can migrate there. Consequently many of the photoelectrons are wasted, and the image centre cannot build up with maximum efficiency.

2.6 Calibration and sensitometry

Before any quantitative information can be obtained from an astronomical photograph, a method of calibration is required which will relate a measurable image parameter to the intensity or brightness of the source. The measured parameter may be the density of the image, its position, or its area, or in some cases two or more parameters may be combined pixel-by-pixel to produce an image profile. Automatic measuring machines provide a scale of machine magnitudes from measurements of more than one parameter. The most direct method of calibration for photographic stellar photometry uses a sequence of stars in the required field for which photoelectric magnitudes are already known. The measured photographic parameters are plotted against the photoelectric magnitudes for the sequence of standard stars, and from this calibration curve may be obtained the magnitudes of the remaining stars. This technique is limited by the availability of a suitable standard sequence in the required area, and by the faintest member of the sequence, which may be several magnitudes brighter than the limit of the photograph. To obtain photoelectric measurements of enough fainter stars requires prohibitive amounts of observing time even on large telescopes.

There are various methods of extending the effective limit of the photoelectric sequence. One technique operates by producing a set of secondary star images, on the same plate as the primary images, where the secondary images are fainter than their primary counterparts by an amount which can be calculated. For example a small prism, described as a sub-beam prism, can be mounted in the path of the incoming beam to deflect a portion of the incident light to form the secondary images. The sub-beam prism is described as a Pickering prism if it is placed at the entrance aperture in the parallel beam of light, or as a Racine prism if it is placed in the converging beam near the focal plane. Depending on the effective aperture of the prism, the secondary images can be made to appear about 3 to 7 magnitudes fainter than the primary images. The magnitude interval between primary and secondary images is determined by the relative areas of the prism and telescope aperture. Alternatively, an objective grating, a full-aperture diffraction grating, may be used, mounted in the incoming parallel beam of the telescope. The intensity difference between the zero and first order images can easily be calculated if the spacing of the grating lines is known.

The use of secondary images to extend a photoelectric standard sequence is shown in Figure 2.14 in which calibration curves P and S are constructed from measurements of the primary and secondary images respectively of the photoelectric standard stars. The curves P and S are

Figure 2.14. A secondary image sequence can be used to extend a photoelectric calibration sequence. The calibration curves P and S are obtained from measures of the primary and secondary images, respectively, of photoelectric standard stars. Translation of standard curve S by δm magnitudes extends curve P and allows calibration of stars up to δm magnitudes fainter than the limiting magnitude of the photoelectric standard sequence. δm is the prism constant of a sub-beam prism, or the grating constant of an objective grating.

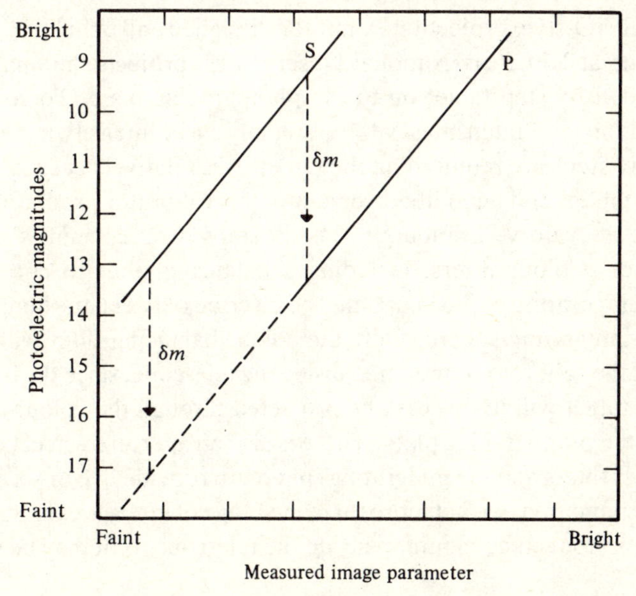

parallel to each other, displaced vertically by an amount δm magnitudes, where δm is the prism or grating constant. If the secondary curve S is moved parallel to the magnitude axis by an amount δm, until the upper part of the secondary curve coincides with the lower end of the primary curve, the calibration of non-standard stars may be extended to a limit δm magnitudes fainter than the faintest member of the photoelectric sequence.

When the calibration is obtained from a sequence of standard stars, the zero point of the magnitude scale is automatically established. If there is no photoelectric sequence available, an alternative magnitude or intensity scale is required, and also a zero point. The relative intensity scale can be obtained from the characteristic curve of the emulsion, as shown below.

To establish the characteristic curve of the emulsion, a calibrator is required which will provide a reference set of light sources, of known relative intensity, which can be recorded on the photographic plate. The simplest calibration device is a neutral-density step wedge, contact-copied onto the emulsion of interest. A more sophisticated and reliable calibrator consists of a stable light source which can be filtered to provide the required colour ranges, an attenuator to provide sufficient steps of different intensities, and a means of imaging the resulting intensity pattern on to the photographic plate. Figure 2.15 outlines three types of calibrator currently in use.

The KPNO-style projection calibrator, designed and built by Hoag and Schoening at Kitt Peak National Observatory, projects the image of a neutral density step tablet on to the photographic plate. To cover the required range of intensities with reasonably small intensity increments, at least 16 steps are required on the tablet. It is relatively easy to change the step tablet, and a modified version of this calibrator produced at the Royal Observatory, Edinburgh, has 25 steps on a 25 mm × 25 mm attenuator. Colour filters, including a balancing filter to correct the colour temperature of the lamp, may be inserted between the lamp house and the integrating sphere. Only the colour balancing filter will be required if the calibrator is mounted inside the telescope, since the image of the step tablet will in this case be projected through the colour filter in front of the photographic plate. There is also an aperture wheel between the colour filters and the integrating sphere, to provide five or six steps of coarse attenuation, so that more than one range of intensities is available. This calibrator can be mounted inside the telescope, or it may be used in

Figure 2.15. Schematic arrangements of calibrators and sensitometers. (a) A projection or step-wedge calibrator. (b) A tube-stack calibrator. (c) A calibration spectrograph. In each case the format of the attenuation component is shown, and its position in the calibrator is indicated.

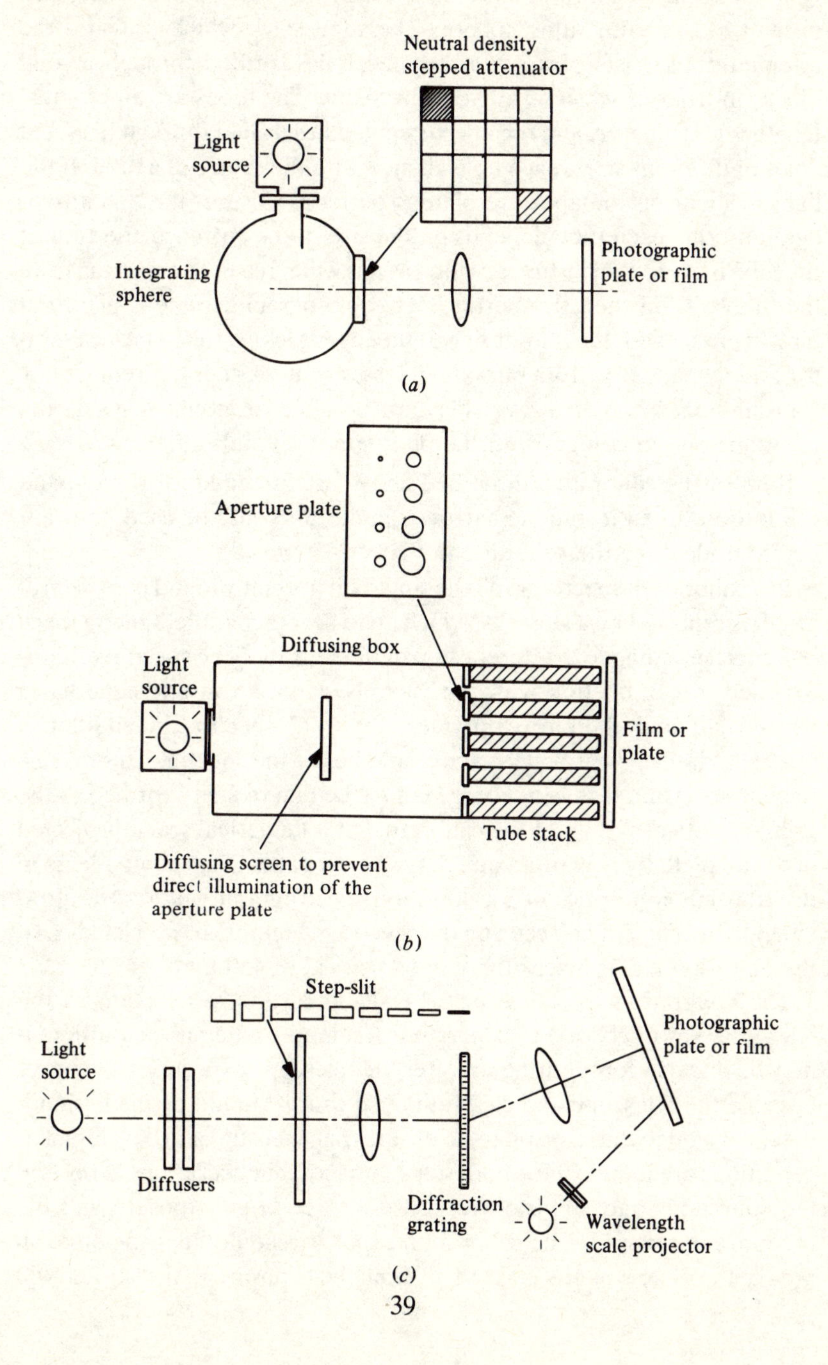

the laboratory. Ultraviolet light may be lost if any optical transmission components are made of glass.

A tube-stack calibrator may have a mirror tunnel or diffusing box instead of the integrating sphere. The tube stack which provides the attenuation consists of a set of parallel holes, drilled through a solid block, in front of which is an aperture plate. The tubes are all of equal length and diameter, and the aperture plate has drilled through it a set of holes of different sizes, each hole aligned with the centre of a tube. If the light incident on the aperture plate is perfectly diffuse, the quantity of light entering each tube depends on the size of the entrance aperture; if the tube is at least three times as long as it is wide, the light emerging from the tube will fill the exit aperture. The photographic plate is placed in direct contact with the output side of the tube stack. A tube-stack calibrator can be made free from ultraviolet losses, since it does not require any transmission optics. However it is not possible to mount it inside the telescope and in contact with the photographic emulsion.

Both of the calibrators described above are intended for broad-band calibration. Even if suitable narrow-band filters could be used, they are far from ideal for the calibration of spectra.

In a calibration spectrograph the single slit normally found in a spectrograph is replaced by a step-slit, which gives several parallel spectra, each one corresponding to a different step of the step-slit. The spectral calibrator used in conjunction with the Isaac Newton 2.5 m telescope has a 12-step slit and therefore provides up to 12 spectra. In addition a wavelength scale is provided, above and below the spectra. This may be the line spectrum from a discharge lamp (see Figure 2.6a), formed by the primary optical system of the calibrator, or a numerical scale, projected onto the plate by way of an auxiliary optical path. Glass components in the transmission optics of a calibration spectrograph will absorb ultraviolet light. The effect is seen on the spectra in Figure 2.6, which look as if the emulsion has lost sensitivity in shortward of 450 nm.

The calibration exposure may be made on the same plate as the telescope exposure, or on a separate test sample of the same emulsion; it may be made before, during, or after the telescope exposure. Whenever possible the telescope and calibration exposures should be made simultaneously, on to the same plate, to ensure that exposure and development conditions are identical for both sets of images, and variations in observing conditions and emulsion response between one exposure and the other are minimised. The calibration exposure should be made on to an area of the plate which is masked off from the exposure to the sky and will therefore be free from contamination by star images. This area should not be too near an original uncut edge of the plate, since the photometric properties of a photographic emulsion are particularly unreliable within 20 or 30 mm of an original edge. Most plates smaller than 250 mm × 200 mm are cut from larger original plates, and should therefore have at least two non-original edges. The UK Schmidt telescope uses two projection calibrators, mounted inside the telescope and coupled to the timing system of the telescope shutter, so that the calibration exposure is exactly simultaneous with the telescope exposure and both exposures are taken through the same filter. If the calibrator cannot be installed inside the telescope, the calibration exposure may still be made on to a masked-off area of the telescope plate, either immediately before or immediately after exposure in the telescope, using a calibrator in the plate loading room near the telescope, or in the laboratory, in conditions as close as possible to those pertaining during the exposure in the telescope. Particular care must be taken to avoid any mismatch between the colour and bandwidth of the telescope and calibration exposures, the exposure times, and any variations in emulsion response due to change in the ambient temperature or humidity. Sometimes it is impossible to put the calibration exposure on the same plate as the telescope exposure – the plate may be too small to accept both exposures, or it may not fit the calibrator. In cases like this the only remaining option is to impress the calibration exposure on to a second photographic plate of the same emulsion type and batch as the first one (both plates preferably being cut from the same original plate) which has been subjected to exactly the same history of handling and treatment. In addition to the difficulties described above, which can arise from sequential exposure, care is needed to ensure that both plates receive exactly the same development.

Before a characteristic curve can be plotted from measurements of a calibration plate, an intensity scale must be established for the calibrator. Approximate intensities can be obtained from measurements of physical parameters of the calibrators. In the case of the tube stack, the relative intensities at the output of the tube stack depend on the relative areas of the corresponding input apertures. For the calibration spectrograph a scale of relative intensities can be found from the widths of the steps in the step-slit, although this does not take into account diffraction effects at the narrowest slits. Similarly an intensity scale of the projection calibrator can be obtained by measuring the diffuse densities of the attenuator steps and calculating the relative transmission of each step. In each case the intensity scales obtained from such measurements should only be taken as first approximations, since no account has so far been taken of other effects, such as diffraction effects or scattered light which may occur in the system between the attenuator and the photographic plate. In order to establish a relative intensity scale which takes all of these complications into account, the output intensities at the plane of the emulsion should be measured individually with a suitable photometer. This is reasonably easy to do for a tube calibrator since the positions of the spots to be recorded are clearly defined by the output face of the tube stack, and the results are usually in good agreement with the first approximation. Direct photometry of the output from the spectral calibrator and the projection calibrators is also highly desirable, although it is more difficult to place the photometer exactly at the image position, and the image itself may be contaminated by the excessive amount of light which can be transmitted by small pinholes in the attenuators.

All of these intensity calibrations are on relative scales, and a zero point is required in each case assuming that there are no stars available with photoelectric magnitudes. Photoelectric measurements of the night sky brightness at the time of the telescope exposure have also been used, in conjunction with the sky fog density recorded on the photographic plate, to provide a magnitude zero point.

Sensitometry is the investigation in the laboratory of the photographic response and its dependence on various parameters. A sensitometer is a device to provide a repeatable set of known relative exposures, covering a range of intensity values of perhaps 1000 to 1 which are impressed on the emulsion under investigation. After processing, the response of the emulsion is evaluated by measuring the densities recorded by the emulsion and relating them to the input intensities by means of a characteristic curve. Clearly sensitometry is very closely related to calibration, and essentially the same devices are used to produce both calibration and sensitometric exposures. The calibrators described above are often referred to as sensitometers, and the two names are often interchanged. There is a continuous requirement for laboratory sensitometric tests at any observatory engaged in regular photographic astronomy, in order to evaluate the response of each batch of emulsion and to establish the optimum exposure which it will require in the telescope. It is not necessary to use a full-size photographic plate for sensitometric tests. Many small plates may be cut from one full-sized plate for use as test samples, though it is essential that the test samples be developed in exactly the same way as the full-sized plates. From measurements of the characteristic curves obtained from such tests, it becomes possible to specify the optimum combination of emulsion, hypersensitisation and development for a particular application at the telescope, and to predict the exposure time needed at the telescope to obtain the required output signal-to-noise from the chosen emulsion. All this can be done independently of test exposures in the telescope, thereby saving valuable telescope time and making optimum use of the telescope for astronomical observations.

Since 1975 members of the American Astronomical Society Working Group on Photographic Materials have been working towards the establishment of a practicable standard of absolute intensity measurements, and methods of obtaining such measures from calibrators and sensitometers. The recommendation is that the measured absolute flux at the emulsion should be expressed in terms of the number of photons per 1000 μm^2 area received in a specified exposure time through a specified narrow-band filter. Results obtained in terms of absolute intensity and standard diffuse density are directly comparable between laboratories using the recommended standards. Such measures also allow the evaluation of absolute values of the detective quantum efficiency of photographic emulsions, and make possible the direct comparison of the DQEs of photographic and electronic detectors.

2.7 Signal-to-noise ratio

The astronomer wants to abstract the information from a photograph as effectively and as easily as possible. The photograph must therefore be taken in order to maximise the output signal-to-noise obtainable from the completed photograph. Although the astronomer has no control over the signal-to-noise of the information emitted by the source he wishes to observe, he does have a small degree of control over the effective signal-to-noise of the total signal incident on the detector, in so far as he can select the focal ratio of the telescope, and exercise some control over the observing conditions.

The output signal-to-noise ratio is a measure of the precision with which the exposure E may be determined and may be written as E/σ_E where E is the exposure and σ_E the rms uncertainty in E. The response of the photographic emulsion is non-linear, and the input exposure is related to the output density as shown in Figure 2.16. It is more helpful to express the output signal-to-noise in terms which can be derived from the normal density versus log E curve. If E and D in Figure 2.16 are corresponding values of exposure and density, and the rms uncertainties in these values are σ_E and σ_D respectively, then over a small part of the D versus E curve $\sigma_E/\sigma_D = \mathrm{d}E/\mathrm{d}D$. Also $\gamma = \mathrm{d}D/\mathrm{d}\log E$. Since

$$\mathrm{d}E/E = \mathrm{d}(\ln E) = \mathrm{d}(\log E)/\log \mathrm{e} = \mathrm{d}(\log E)/0.4343,$$

$$\text{output signal-to-noise} = \frac{E}{\sigma_E} = \frac{E}{\sigma_D}\frac{\mathrm{d}D}{\mathrm{d}E} = \frac{0.4343}{\sigma_D}\frac{\mathrm{d}D}{\mathrm{d}\log E}$$

$$= \frac{0.4343\gamma}{\sigma_D}$$

In this relation, γ is the local value of the gamma at the point on the density–log E curve corresponding to the density D and exposure E in Figure 2.16. It has been shown that the density noise σ_D depends only on the grain type of the emulsion and the absolute or total diffuse density – that is, the sum of the gross fog density, sky fog density, and the signal density. It is independent of hypersensitising treatment and of the colour sensitivity of the emulsion. The zero point of the density scale is established by removing the sample from the measuring beam of the densitometer. Values of σ_D as a function of absolute density for several emulsions are given in Table 2.1. As may be expected, the density noise is greater for

Figure 2.16. The expression relating output signal-to-noise, as derived in Section 2.7, is obtained from the density versus exposure relation for a given emulsion type.

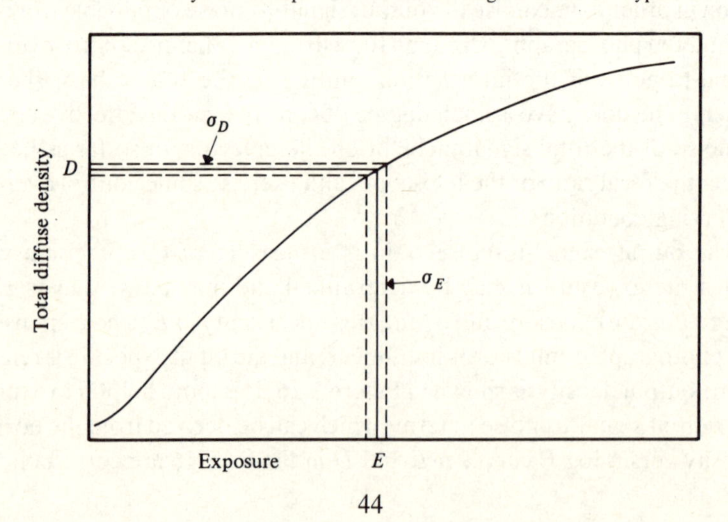

Table 2.1. The density noise σ_D depends on the emulsion type, its total diffuse density D and the size of the aperture with which it is measured. The values of σ_D in this table were obtained using an aperture of 1000 μm^2. (From Furenlid, in West & Heudier, 1978: p.158.)

Total density	σ_D						
	103a-O	IIa-O	IIIa-J	IIa-D	IIa-F	098-04	IIIa-F
0.2	–	0.017	0.007	0.020	0.017	0.017	0.007
0.4	0.037	0.023	0.011	0.025	0.023	0.026	0.011
0.6	0.039	0.028	0.014	0.030	0.028	0.032	0.015
0.8	0.041	0.033	0.017	0.035	0.033	0.037	0.017
1.0	0.046	0.038	0.019	0.039	0.038	0.041	0.018
1.2	0.051	0.042	0.022	0.044	0.043	0.048	0.021
1.5	0.060	0.048	0.025	0.053	0.050	0.058	0.026
2.0	0.079	0.058	0.031	0.066	0.061	0.076	0.034
2.5	0.101	0.068	0.036	0.075	0.069	0.094	0.039
3.0	0.129	0.075	0.038	0.083	0.077	0.113	0.044

coarse-grained emulsions than for fine-grain emulsions. The expression derived above shows that to be capable of a high peak value of output signal-to-noise an emulsion must have high contrast and fine grain.

The evaluation of the output signal-to-noise behaviour of a given emulsion as a function of density or of exposure is now relatively simple. The characteristic curve is constructed from measures of a series of calibration or sensitometric spots and the output signal-to-noise calculated from the above equation, the data in Table 2.1 and the local values of γ obtained from the characteristic curve at the appropriate density levels. The dependence of output signal-to-noise on density is typically of the form shown in Figure 2.17. For most spectroscopic emulsions the peak value of output signal-to-noise corresponds to an exposure level which gives a total density of about 1.0 (diffuse), and the loss of output

Figure 2.17. The dependence of output signal-to-noise on total density, for two different emulsions. The density levels at which the output signal-to-noise falls to 75 per cent of its maximum value are indicated by bars on the curves.

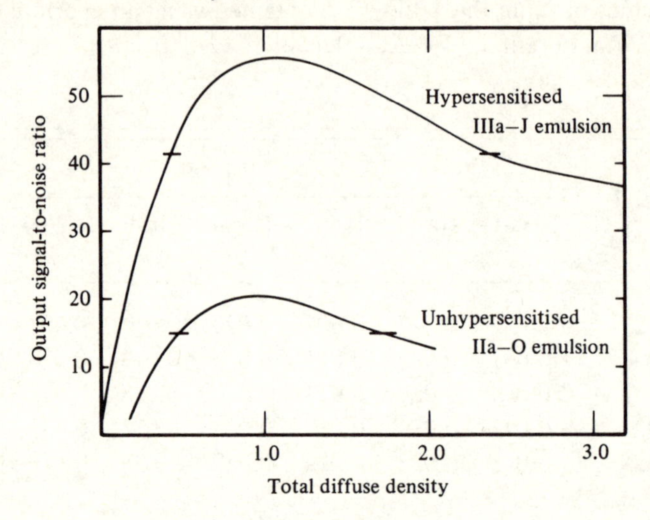

signal-to-noise is greater if the emulsion is underexposed than if it is overexposed. Independent visual inspection of test plates supports these results, but strictly the present discussion refers to the evaluation of output signal-to-noise from density-recording microphotometer measurements.

The signal-to-noise ratio of the input signal improves with exposure or integration time. A signal E has an associated photon noise component equal to $E/E^{1/2}$ or simply $E^{1/2}$.

In Chapter 1 detective quantum efficiency and signal-to-noise were related by the expression

$$\mathrm{DQE} = (S/N_{\mathrm{out}})^2/(S/N_{\mathrm{in}})^2$$

Substituting for the signal-to-noise ratios gives

$$\mathrm{DQE} = (0.4343)^2 \times \frac{\gamma^2}{\sigma_D^2} \times \frac{1}{E} = \frac{0.1886}{E} \times \frac{\gamma^2}{\sigma_D^2}$$

In Section 2.2 it was shown that if E is the exposure corresponding to the required output density D, then E may be considered to be a measure of the speed of the emulsion. An emulsion capable of operating at high peak DQE will therefore have high contrast and fine grain, since this relation is dominated by the squared terms, rather than high speed. If the exposure value E can be evaluated in absolute terms of photons per unit area received in the specified exposure time, then the absolute DQE can be obtained quite simply, permitting direct comparison with other types of detectors. Figure 2.18 shows typical curves of density, input signal-to-noise and DQE as a function of log E. It can be seen that the peak value of DQE corresponds to a lower value of E than does the peak output signal-to-noise. Since to the astronomer the quality of the output is more important than the efficiency of operation of the detector, the emulsion should be operated so as to maximise the output signal-to-noise ratio rather than the DQE. Figure 2.18 also shows that the peak value of the output signal-to-noise ratio occurs at the same exposure level as the lowest point of the straight-line portion of the D versus log E curve, where gamma first reaches its highest value. This is also at the higher of the two values of E at which the input signal-to-noise and DQE curves intersect.

Figure 2.18. The variation of input signal-to-noise, output signal-to-noise, DQE and final density with relative log exposure, from a sample of type IIa emulsion. Input signal-to-noise is plotted on an arbitrary scale, and in the absence of absolute exposure values the DQE curves are also on a relative scale.

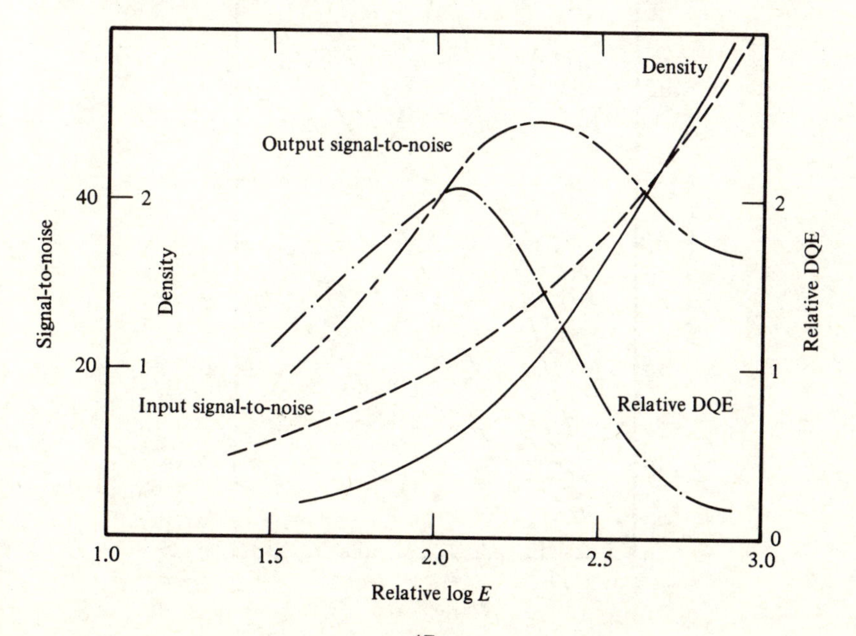

Consider the emulsions whose characteristic curves are shown in Figure 2.19. The shape of the characteristic curves implies that the IIa-O will record those signals which give densities below 1.8 with less exposure than would be required by the IIIa-J. However, the corresponding output signal-to-noise curves show that except for the very weak signals, the IIIa-J emulsion will give more precise information than the IIa-O over most of its exposure range.

The detection of weak signals may be in conditions of high input signal-to-noise ratio, sometimes referred to as Class I detection, or in conditions of low input signal-to-noise ratio, referred to as Class II detection.

Class I detection conditions occur when the signal is strong, but also when a weak signal is accompanied by very little noise. Examples of these conditions are when working at long focal ratios, or at shorter foci if interference filters are used which effectively eliminate the sky background noise. In this case the recording situation is signal limited. To

Figure 2.19. A comparison of the output signal-to-noise response of a IIIa-J emulsion and a IIa-O emulsion of approximately comparable speed. Both emulsions have been hypersensitised. The output signal-to-noise of the IIa-O emulsion is only above that of the IIIa-J emulsion for very low exposures.

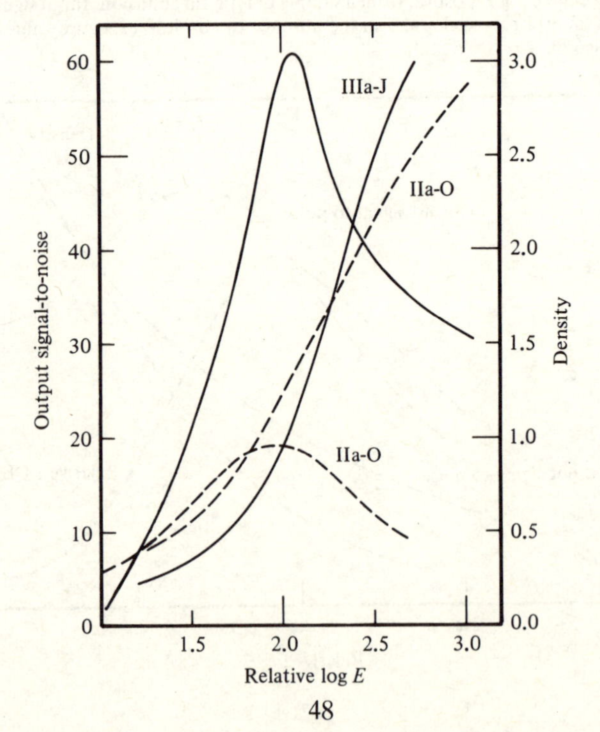

optimise the output signal-to-noise from the emulsion, the exposure should be long enough to record the signal at the lowest part of the straight-line portion of the H & D curve. If there is enough exposure available to record the image at a higher density, a better value of output signal-to-noise could be obtained by using a different emulsion, with a finer grain structure, despite the lower sensitivity of the fine-grain material. If the input signal is too weak or the only available emulsion is too insensitive to permit the image to be recorded at the optimum point of the characteristic curve, the total exposure can be raised to the optimum point by giving the emulsion a preliminary exposure to a light source of uniform intensity. This additional exposure or preflash raises the operating level of the emulsion to a higher point of the H & D curve, where the DQE and the output signal-to-noise are both improved. The level of the preflash exposure must be limited so that the total density due to the signal exposure, preflash and gross fog will not exceed the density level corresponding to the maximum output signal-to-noise. The optimum level of preflash is usually one which gives a total flash plus gross fog density of 0.3.

In Class II detection conditions the input signal has to be recorded in the presence of a large amount of noise. Such conditions are most commonly found in broad-band photometry of faint objects, especially with fast Schmidt telescopes where exposures are often limited by the brightness of the sky. Another example is the detection of faint structural details in extended sources such as galaxies or nebulae. Frequently the signal consists of fewer photons than are contributed by the sky background and the operating level of the emulsion is dominated by the background noise rather than by the signal level. Since the input signal-to-noise is already poor, the detecting emulsion must degrade the input signal-to-noise as little as possible, that is, it must be capable of operating at high DQE. The exposure level of the input signal plus noise is only fractionally higher than that of the background noise alone, and to obtain a large difference in output density between the two requires high contrast emulsion. Emulsions which satisfy these requirements are usually very slow, and suffer from low intensity reciprocity failure. However they generally respond well to hypersensitisation. The type IIIa emulsions were specially designed for operation in Class II situations, and are capable of operating at a peak DQE of about 4 per cent.

References for Chapter 2

Dainty, J.C. & Shaw, R., 1974. *Image Science*, Academic Press.

Eastman Kodak publication no P-315. Plates and films for scientific photography, 1973.

Heudier, J.L. & Sim, M.E. (eds.), 1981. *Astronomical Photography 1981* – Proceedings of meeting of International Astronomical Union Working Group on Photographic Problems held in Nice, April 1981.

James, T.H. (ed.), 1977. *The Theory of the Photographic Process*, 4th edition, Macmillan Company.

West, R.M. & Heudier, J.L. (eds.), 1978. *Modern Techniques in Astronomical Photography* – Proceedings of ESO-sponsored Workshop held in Geneva, May 1978.

3

Current photographic techniques in astronomy

The techniques currently employed in astronomical photography have evolved rather rapidly in recent years, partly as a result of the introduction of new materials, and partly because of the need to systematise and optimise handling and processing methods at observatories where the photographic throughput is high. Some dedicated photographic telescopes employ specialist staff to handle all apects of the preparation, observation and processing, without the astronomer for whom the plates are intended needing to be present at all, on the grounds that the full-time specialist is more efficient than the visitor who may only spend a few nights at a time at the telescope. Many of the techniques to be described in this chapter have been developed at observatories like these.

3.1 Available materials

The range of commercially available emulsions capable of responding to very faint light levels is dominated by the spectroscopic emulsions made by the Eastman Kodak Company and consequently this section will be restricted to considerations of available Kodak products. Certain emulsions are also produced by ORWO in East Germany and by NII Techphotoproject in Kazan in the USSR. Although these products are limited in availability and range, they appear to include equivalents of the Kodak IIIa-J and IIa-O emulsions.

The group of Kodak spectroscopic emulsions which include 'a' in their type reference have been specially treated in manufacture to reduce low intensity reciprocity failure. The type reference also includes indications of the grain class and spectral response of the emulsion. The sequence I, 103, II, III, IV is approximately one of increasing gamma, dynamic range and DQE, and of decreasing speed and grain size. The sequence O, J, G, D, E, F, N, Z is in increasing order of long-wavelength sensitivity. Hence emulsion type IIIa-J is a very fine-grain emulsion, with a long wavelength cutoff in the blue–green region, which has been treated in manufacture to reduce low intensity reciprocity failure.

The natural sensitivity of all emulsions is in the ultraviolet and blue region of the spectrum. The sensitivity of emulsions 103a-O and IIa-O corresponds to these regions. In order to extend the colour sensitivity to longer wavelengths, suitable sensitising dyes are incorporated into the emulsion during manufacture. The sensitivity to blue and ultraviolet light is retained in the dyed emulsion, although dyed emulsions will normally be used to record radiation in the dye-sensitised region. Examples of the natural and dyed sensitivities are shown in Figure 3.1 and in Figure 3.2, which also indicates the range of colour sensitisings available to astronomers. If a limited range of wavelength sensitivity is required – for instance, a region in red light to include H α – the long wavelength limit is usually defined by the long wavelength cutoff of the emulsion sensitivity, and the short wavelength limit by a colour filter which blocks the unwanted short wavelengths. This is illustrated in Figure 3.3.

Not all combinations of colour sensitivity and grain class are available.

Figure 3.1. The natural sensitivity of the photographic emulsion is to blue and ultraviolet light. The sensitivity can be extended towards the red by the addition of appropriate dye sensitisers during manufacture.

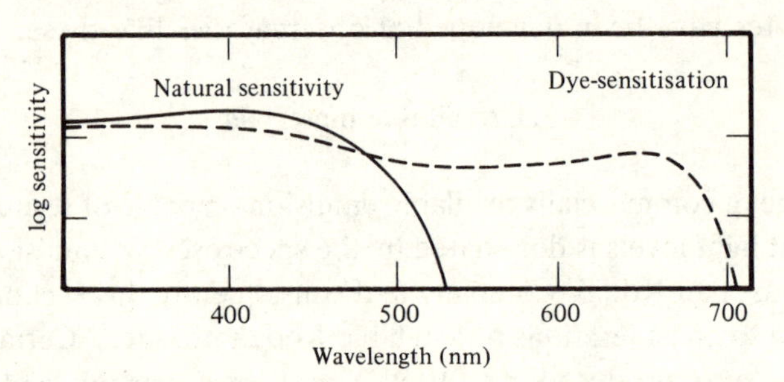

Figure 3.2. A summary of the spectral sensitivity classes and corresponding class designation letter for currently available Kodak spectroscopic emulsions. The shaded portions indicate particularly useful wavelength ranges. The spectral range of 098 emulsion and Technical Pan 2415 are the same as are shown for class F emulsions.

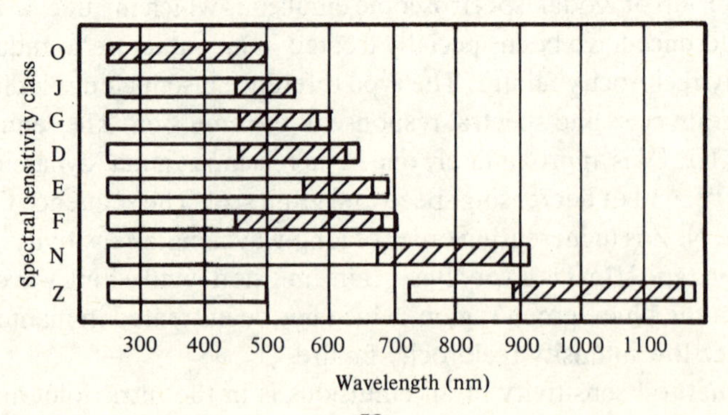

Figure 3.3. The wavelength range required for an observation can be obtained by selecting an emulsion whose sensitivity cutoff defines the long wavelength limit and combining it with a filter to cut out the shorter wavelengths.

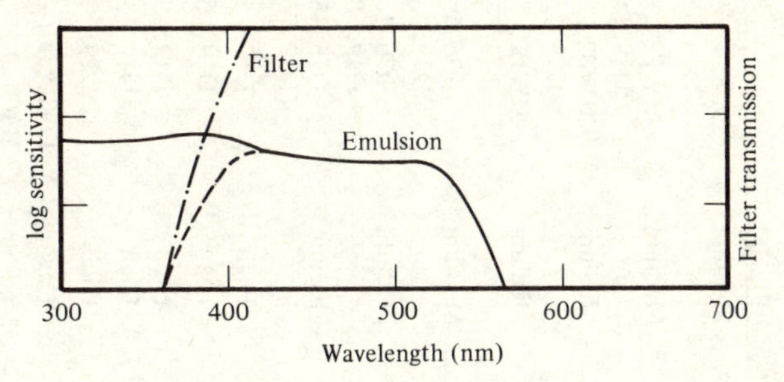

Table 3.1 shows some properties of the combinations most commonly used in astronomy, available with the emulsion coated onto glass plates. Some but not all of these are also available on film. The long wavelength limit indicates the effective working limit, taking into account the steep fall in sensitivity at the long wavelengths. Emulsions of any type will vary slightly from batch-to-batch and with development procedures. Consequently the user always needs to test each batch of emulsion in his own laboratory to obtain information representative of his system.

Granularity is defined as the rms deviation in diffuse density when a sample of exposed and processed emulsion having a specified diffuse density is measured with a specified aperture. Granularity is an objective measure of the effect of grain noise, and should not be confused with graininess, which is a subjective assessment of the visual effects of grain in a photograph. Since the measured granularity of an emulsion depends on both the size of the measuring aperture and the density of the emulsion, both of these parameters must be specified before comparisons can be made. The figures quoted in Table 3.1 refer to measurements at 1.0 total diffuse density obtained with an aperture of area 1000 μm^2, and are obtained from Table 2.1, or derived from the published Kodak information. It can be seen from the table that emulsions in classes III and IV show significantly smaller values of granularity than those in classes I, II and 103.

The resolving power of an emulsion is a measure of its ability to maintain the separate identity of images formed by closely adjacent objects. As explained in Section 1.4, this can be expressed in terms of the number of parallel lines per millimetre which can be recorded by the emulsion whilst retaining distinct separation of the images. The ability of an emulsion to resolve a sinusoidal intensity distribution is described by

Table 3.1. Some properties of Kodak emulsions used in astronomical photography and suggested developers for them.

Emulsion	Long wave limit (nm)	RMS granularity (1000 μm^2 area)	Resolving power (line-pairs per mm)	Contrast class	Developer
103a-O	500	0.046	80	Medium	MWP2, 9 mins D19, 5 mins
103a-E	660	0.046	80	Medium	MWP2, 9 mins D19, 5 mins
103a-F	680	0.046	80	Medium	MWP2, 9 mins D19, 5 mins
098	680	0.041	63	Medium	MWP2, 9 mins D19, 5 mins
IIa-O	500	0.038	87	Medium	D76, 15 mins D19, 5 mins
IIa-D	630	0.039	87	Medium	D76, 15 mins D19, 5 mins
IIa-F	680	0.038	100	Medium	D76, 15 mins D19, 5 mins
IIIa-J	550	0.019	200	High	D19, 5 mins
IIIa-F	680	0.018	200	High	D19, 5 mins
I-N	890	0.044	100	High	D19, 5 mins
IV-N	890	0.022	200	High	D19, 5 mins
I-Z	1160	0.040	125	High	D19, 5 mins
Kodak	680	0.009	400	Low	POTA, 10–20 mins
Technical		0.011	320	Medium	HC110-D, 8 mins
Pan Film 2415		0.010		High	D19, 5 mins

its modulation transfer function (MTF). Examples of MTF curves are shown in Figure 3.4. MTF curves for each component of an optical and photographic system can be combined very easily to give the overall MTF of the system.

It is not possible to relate the MTF directly to the resolving power of an emulsion since resolving power depends on the granularity and small-scale sensitivity of the emulsion as well as on its MTF. The sequence I, 103, II, III, IV is only approximately related to granularity, since the range of grain sizes within an emulsion is significant, as well as the mean grain size.

When selecting an emulsion for a particular application, the required resolving power of the emulsion is determined by the image resolution in the final focal plane. In direct imaging, the angular resolution of the image is determined by the seeing and the resolution of the telescope optics. In a spectrograph the spatial resolution of the image is normally determined entirely by the size of the spectrograph slit and the geometry of the camera. In either case the chosen emulsion must be capable of resolving all the spatial frequencies present in the image.

Most emulsions are available coated onto glass plates, and several are available on a film base. In either case the back of the plate or film may be coated with a special antihalation backing, to reduce or eliminate the

Figure 3.4. Typical modulation transfer curves for two different emulsions. Emulsion A records lower frequencies more effectively than does emulsion B, but at frequencies above 30 cycles per mm emulsion B performs better than emulsion A.

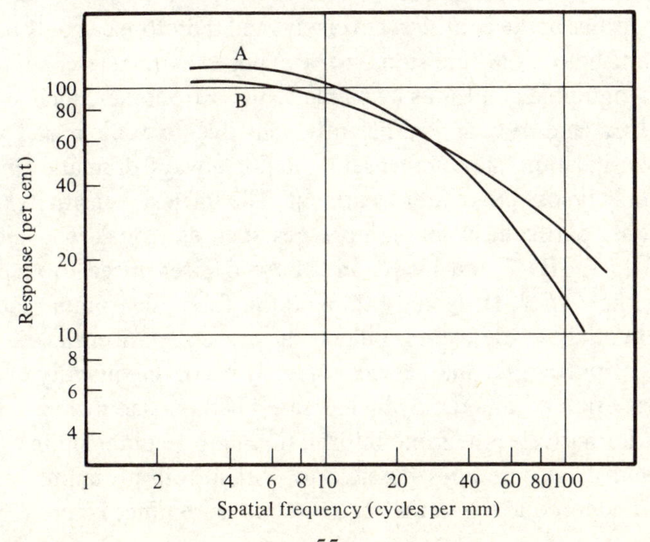

intensity of the halation ring found round the images of bright stars. These rings occur when light in the main image is internally reflected off the back surface of the glass or film base and back into the emulsion. The backing is designed to dissolve in the developer, and is usually available in standard or high density forms. It is usual to select high density antihalation backing for most spectral sensitisings, but infrared plates are often used without backing, since even the high density backing is relatively ineffective at these wavelengths, and it also has a tendency to come off in patches during the hypersensitisation procedures.

A promising recent addition to the range of materials available to the photographic astronomer is Kodak Technical Pan Film 2415, previously known as SO-115. This emulsion, available only on film at the time of writing, has extended panchromatic wavelength sensitivity, extremely fine grain, and can be processed to give gamma values from 0.4 to over 3, depending on the developer used. For astronomical applications the 2415 emulsion must be hypersensitised. Further details about this emulsion will be found in the American Astronomical Society (AAS) Photobulletin and in the appropriate Kodak publications.

3.2 Hypersensitisation

Hypersensitisation is the subjection of the unexposed emulsion to one or more methods of physical or chemical treatment in order to improve its response to light. Most methods of treatment affect the long-term keeping properties of the emulsion adversely, and therefore can only be used a short time before the emulsion is to be exposed in the telescope. Some of the photographic emulsions available to the astronomer, such as Kodak types 103a and IIa, are sufficiently sensitive to be exposed without hypersensitisation, although useful, if not always dramatic, improvements may be obtained after treatment. The intrinsic sensitivity of other emulsions, particularly fine-grain types such as Kodak Technical Pan 2415, IIIa-J, IIIa-F and IV-N, and those dye-sensitised to respond to infrared light (I-N, IV-N and I-Z) is so low that they cannot reach their full potential as detectors unless they are hypersensitised. When hypersensitisation techniques are correctly used, the quantity or quality of information obtained may be increased without sacrifice of telescope time. Alternatively, the same information may be obtained in a shorter time, and it is sometimes possible to obtain both an improvement in quality and a reduction in the required exposure time. Figure 3.5 shows

Figure 3.5. Output signal-to-noise ratio and characteristic curves for hypersensitised and untreated samples of IIa-O emulsion. Both samples were cut from the same original plate.

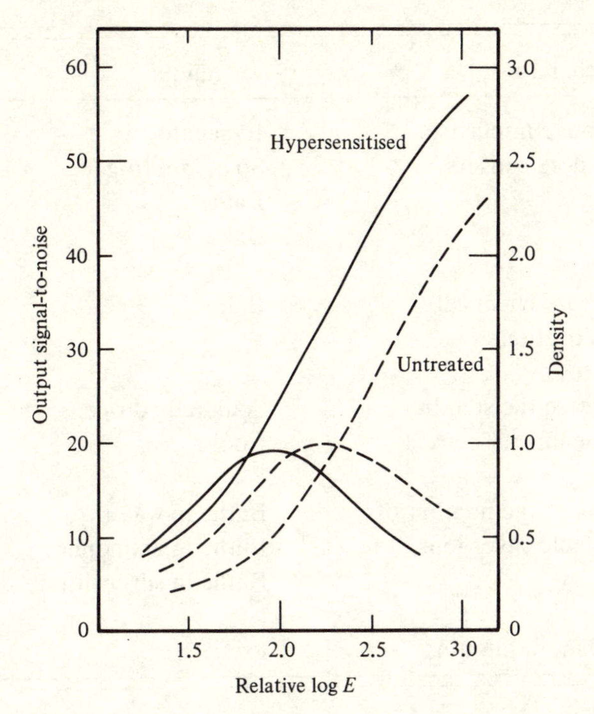

the characteristic curves and signal-to-noise responses of untreated and hypersensitised samples of IIa-O emulsion, both from the same batch. The hypersensitised sample is about 3 times faster than the untreated emulsion, in terms of the exposure time required to reach a density of 0.6 diffuse above chemical fog, and the peak values of the signal-to-noise curves indicate that any loss of quality of information due to hypersensitisation is minimal.

There are several techniques of hypersensitisation in regular use at observatories throughout the world, but the mechanisms by which some of them work are not yet fully understood. Table 3.2 summarises current opinion on the mechanisms believed to be responsible for improvements due to most of those techniques. Techniques which work well on some emulsions may have disastrous effects on others. Table 3.3 indicates those combinations of emulsion and hypersensitising method which have proved useful or otherwise. It is now known that oxygen and water vapour, together or separately in an emulsion, act as desensitisers. They may be removed by storing the emulsion in an atmosphere of pure dry

Table 3.2. Mechanisms of hypersensitisation, and laboratory hypersensitising techniques which utilise them.

	Mechanism	Technique
1.	Remove impurities and desensitisers	Evacuate Soak in nitrogen Bathe Bake
2.	Increase chemical sensitisation	Bake
3.	Increase the stability of the image speck	Soak in hydrogen Cool
4.	Increase the number of available silver ions	Bathe in water Bathe in ammonia Bathe in silver nitrate
5.	Add more photons	Preflash

nitrogen, or by placing the emulsion in a suitable tank and evacuating it. However, if the vacuum is too severe, it may cause the surface of the emulsion to harden and become impermeable, trapping oxygen and water vapour deeper in the emulsion. Raising the temperature not only increases the rate at which the oxygen and water vapour can diffuse from the emulsion, but it may also extend some of the chemical sensitisation processes which were begun in manufacture. Hypersensitisation by baking has been a relatively popular procedure for many years, although baking in air has now been discarded in favour of baking in nitrogen, which gives more uniform results than baking in air and is easier than baking in vacuum.

A significant advance in hypersensitisation methods began in 1974 with the discovery that hydrogen gas is a very effective hypersensitising agent. An emulsion stored for a few hours in hydrogen gas at room temperature showed improvements in performance at least as great as those achieved by the same emulsion after several hours of baking in dry nitrogen. The effect of hydrogen treatment on type IIa-O emulsion is shown in Figure 3.6. It is now believed that the hydrogen acts as a reduction sensitiser,

Table 3.3. A summary of the effectiveness of various hypersensitising methods when applied to different emulsion types. Key: VE – very effective; E – effective and useful; e – only a small effect, not particularly useful; X – to be avoided; a space indicates this method was not tested for the particular combination and *italic* indicates the method is preferred by at least one user.

Category of use		Class I (high $S{:}N_{in}$, signal dominated) and Class II (low $S{:}N_{in}$, sky limited)										Class I only
Procedure		Evacuate	Cool	Soak in nitrogen	Bake in air	Bake in nitrogen	Bake in forming gas	Soak in hydrogen	Bathe in water	Bathe in NH_4OH	Bathe in $AgNO_3$	Preflash
Emulsion	103a-O	E		E	e	*E*	*E*	*E*				E
	103a-D					E						
	103a-E			e		E		E				
	103a-F											
	098	E		E		E		E	E	e		E
	IIa-O	E		E	e	*E*	*VE*	*E*				E
	IIa-D			E		E		E				
	IIa-E											
	IIa-F	E	E			E		E				
	IIIa-J	E	E	E	e	*VE*	*VE*	*VE*	X			E
	IIIa-F	E		E	e	E	E	*VE*	X			
Kodak Technical Pan	2415	e		e			E	E			E	
	IV-N	E		e	e	e	E	E	E	*E*	*VE*	E
	I-N	E	E				E		E	*E*	*VE*	
	I-Z								E	*E*	*VE*	
Mechanism												
Further chemical sensitisation					0	0	0					
Reduction sensitising					0	0	0	0	0?	0?	0	
Remove oxygen		0		0		0	0	0				
Remove water vapour		0		0		0	0	0				
Increase Ag^+ concentration									0	0	0	
Reduce recombination			0									
Increase Ag^+ stability			0									
Raise faint image to higher density												0

Figure 3.6. Part of the cluster of galaxies Abell 1060, photographed on IIa-O emulsion. Photograph (a) was taken on an unhypersensitised plate, in 3 hours; photograph (b) was an exposure of 2½ hours onto a plate which had been soaked in hydrogen for 5 hours. Both plates were taken on the same night with the 1.0 metre telescope at Siding Spring Observatory, Australia.

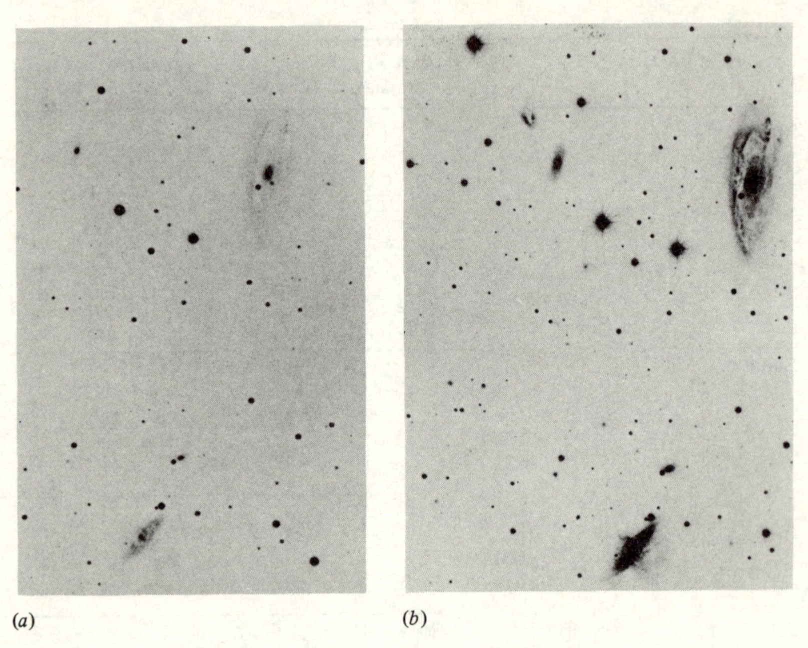

(*a*) (*b*)

assisting the reduction of silver ions to silver atoms and diminishing the recombination of photoelectrons which is largely responsible for low intensity reciprocity failure. The obvious disadvantage of hydrogen is its potentially explosive nature, to the extent that its use may be impracticable on grounds of safety. Indeed it has been banned at some observatories. Two solutions have been found to this problem. At the UK 1.2 metre Schmidt telescope at Siding Spring a special hydrogen hypersensitising unit has been designed and built. The plates to be hypersensitised are loaded into gas-tight boxes which are filled with dry nitrogen and then placed in a special temperature-controlled cupboard outside the main building and connected to gas supply lines. An automatic control system delivers a further purge of nitrogen, and then enough hydrogen to fill the box. At the end of the required soaking time, a further purge of nitrogen is used to remove the hydrogen which is safely vented into the atmosphere. To prevent sparking, the valves which control the hydrogen flow are in turn controlled by nitrogen pressure. The second solution, which is in more general use, is to bake the emulsions in forming gas. This is a non-explosive mixture of hydrogen in nitrogen, which may contain 2 per cent, 5 per cent or 8 per cent of hydrogen. In so far as such comparisons can be made without absolutely calibrated sensitometry, the results achieved by each of these techniques appear to be very similar, and both have been used to great effect on IIIa-J and IIIa-F plates, including those used for the European Southern Observatory/Science Research Council survey of the southern sky. Kodak Technical Pan Film 2415 also responds well to baking in forming gas, or to hydrogen soaking or even to bathing in silver nitrate.

The techniques described so far are particularly applicable to emulsions which are sensitive to visible light, though some of them may also be applicable to infrared sensitive emulsions. The most effective methods of hypersensitising infrared sensitive emulsions involve washing the emulsion in pure water or in a weak solution of silver nitrate or ammonia. Restrainers, usually an excess of bromide ions, are incorporated into the emulsion during manufacture, to give them a reasonable shelf-life. If these restrainers are washed out of the emulsion, the number of available silver ions is increased, although the treated emulsion must be exposed and processed within hours of treatment. Bathing the emulsion in an ammonium solution or a silver nitrate solution increases the number of available ions still further. The spectacular gains which can be achieved by treating the IV-N emulsion in silver nitrate solution have made it possible for the UK Schmidt telescope to undertake a deep photographic survey of the southern galactic plane to a limiting magnitude of about 19 magnitudes in the waveband 715 nm to 900 nm, with exposure times of around 90 minutes per plate. Figure 3.7 shows the capability of IV-N emulsion treated with silver nitrate solution.

The I-Z emulsion is capable of recording signals at 1.1 μm, but it is very slow indeed and cannot be used unless it is hypersensitised. It can be treated in a similar way to the I-N and IV-N emulsions but the most successful method is soaking in silver nitrate. The optimum solution strength required is weaker than that required by the IV-N emulsion, and best results are obtained if the solutions are chilled.

Preflashing, that is the brief exposure of the emulsion to a uniform light source before it receives the main exposure, may improve the emulsion response by providing enough photons to produce sub-latent image specks of silver which contain almost enough silver to be stable and developable. This allows the weak signals in the main exposure to be recorded more effectively, by building on the foundation of the pre-

Figure 3.7. Comparison of photographs taken in the near infrared. Exposure times were all 90 minutes. The upper pair were taken on untreated I-N emulsion, and the lower pair on IV-N emulsion hypersensitised in silver nitrate solution. The barred spiral galaxy on the left is NGC 1365; NGC 1350 is on the right.

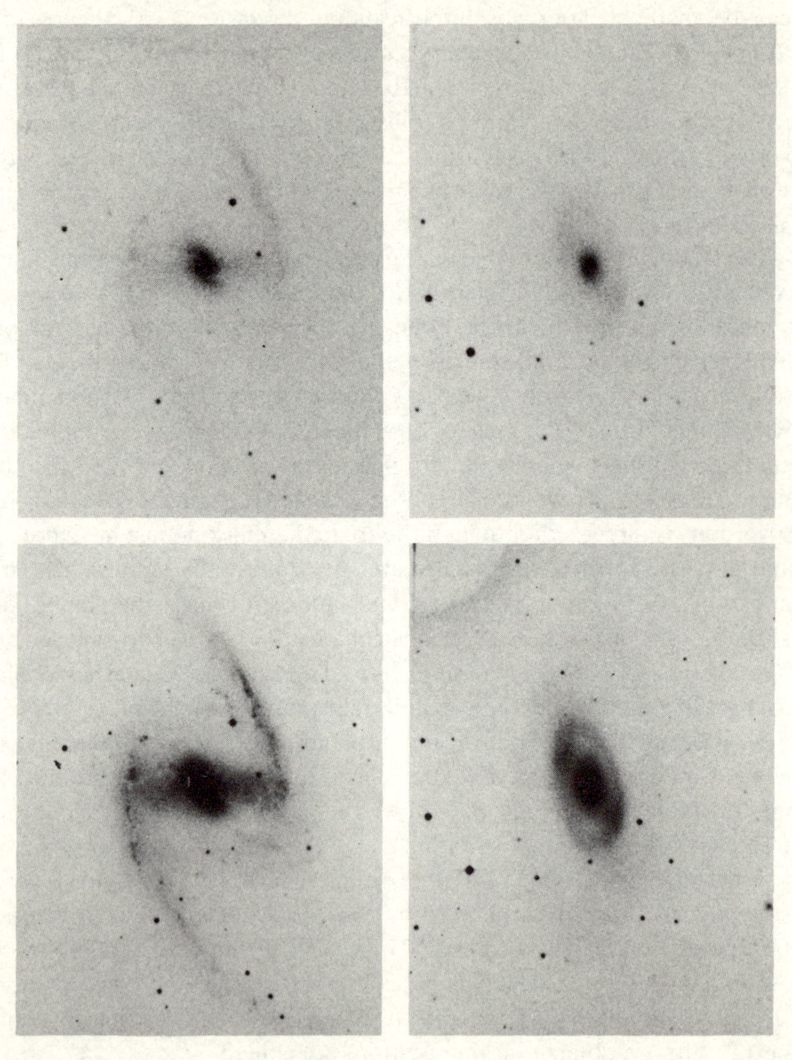

formed latent image centres. Preflashing is only applicable in conditions of high input signal-to-noise detection, or when the input noise, for example the sky background, is extremely low. The additional background density caused by the preflash should be only about 0.2 diffuse density units for optimum effect, and the total density due to chemical fog and preflash fog should not exceed 0.3.

Cooling an emulsion during exposure may assist latent image formation by reducing the mobility of the positive holes and the consequent loss of photoelectrons.

It is often possible to combine sequentially two or more hypersensitising treatments which operate by different mechanisms, to obtain additional increases in sensitivity. Vacuum treatment or nitrogen soaking often precedes baking or hydrogen treatment. Not all combinations are successful, since some lead to excessive fog levels. The most likely combinations to give worthwhile increases in sensitivity without excessive fog levels are those which sequentially use two different mechanisms in the order shown in Table 3.2. For instance, evacuation followed by hydrogen soaking is a useful combination, to which preflashing could be added, but nitrogen soaking after evacuation is fruitless.

The conditions under which hypersensitised emulsions are stored between treatment and exposure need to be very carefully controlled, since the treatment will have reduced their storage life considerably. For most emulsions, storage in dry nitrogen in a refrigerator or at deep freeze temperatures is preferred, although infrared emulsions are better stored in more humid conditions.

The reader contemplating the application of hypersensitising methods or the establishment of a hypersensitising system is advised to consult the technical literature (especially the publications of the International Astronomical Union Working Group on Photographic Problems, and the American Astronomical Society Photobulletins) for details of equipment and procedures currently in use. Hypersensitisation procedures must be used with great attention to fine details, since some methods, especially those involving infrared emulsions, are extremely difficult to control. There is no perfect system which works equally well at all observatories, and no perfect recipe which suits all emulsion types. Consequently each observatory has had to establish and evolve its own particular system to match its own requirements.

A prerequisite for successful hypersensitising is a reliable sensitometric testing facility, since the batch-to-batch variations of emulsion characteristics mean that every batch must be tested to establish its optimum hypersensitisation recipe. A series of sample plates is subjected to progressive degrees of treatment with the hypersensitisation technique under investigation, and then each sample is given exactly the same exposure in a sensitometer and processed in exactly the same conditions as the plates from the telescope. The sample plates are then measured in a densitometer to establish their characteristic curves. Of the several parameters which may be used to evaluate the effects of hypersensitisation, the output signal-to-noise is the one which should be optimised for nearly all applications, since it determines the precision with which the information stored in the detector may be extracted. Figure 3.8 shows the dependence of output signal-to-noise on the hydrogen treatment time for a batch of type IIIa emulsion, expressed as plots of output signal-to-noise against log E. The shift of the signal-to-noise peak towards smaller values of log E indicates an increase in the emulsion speed, and the decrease in the peak output signal-to-noise beyond one hour of hydrogen treatment with no change in the emulsion speed indicates that excessive treatment is not only wasted but harmful. Therefore for optimal recording of faint images on this emulsion the treatment should not exceed two hours in hydrogen. Figure 3.8 also shows that the output signal-to-noise is at its maximum for the untreated emulsion, and decreases with treatment. Consequently if the object to be photographed is sufficiently bright, it can be recorded on an untreated emulsion with better output signal-to-noise than would be obtained from a hypersensitised emulsion of the same type and batch. The interdependence of output signal-to-noise, emulsion type and the hypersensitisation treatment must therefore be taken into account when

Figure 3.8. The effect of hypersensitisation on the output signal-to-noise of type IIIa-F emulsion. The time in hours for which each sample was treated in hydrogen is indicated. The characteristic curve for the untreated emulsion is also shown.

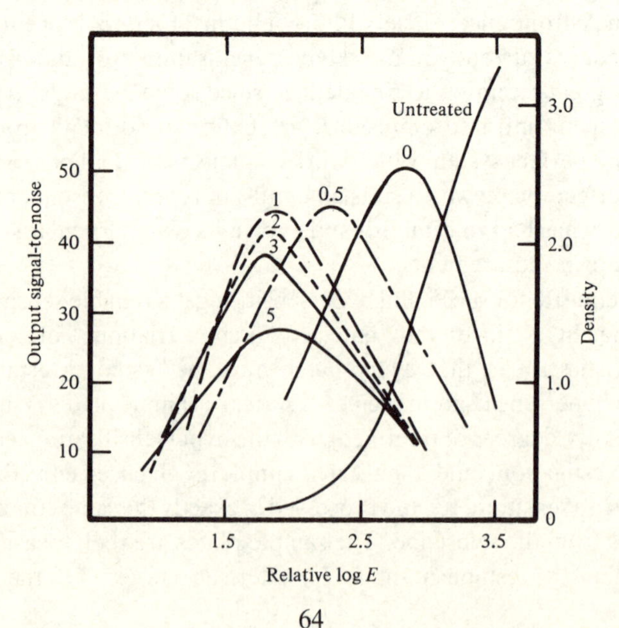

selecting the most appropriate emulsion for a particular application, and deciding whether or not hypersensitisation is required.

3.3 Processing for uniformity and permanence

The difficulty of processing photographic plates to high standards of uniformity increases with the area of the plate. The materials used for photography at the telescope range in size from plates and films only a few millimetres square, to the large plates up to 360 mm square used in some Schmidt telescopes, or the 500 mm square plates used at the Cassegrain focus of the 2.5 m Dupont telescope. Development methods which are optimum for some plate sizes are inappropriate or completely inadequate for other sizes.

The processing of photographic emulsions involves two basic stages. First the exposed emulsion is developed, or chemically amplified, to increase the amount of image silver in exposed grains and so form a visible image. Next the remaining unexposed silver halide is dissolved by fixation, and the dissolved halide and other unwanted byproducts of the chemical processes are removed by washing. Development is the only one of these processes which is not taken to chemical completion, and therefore requires the most attention to factors such as solution strength and temperature, and development time. In order to produce a visible image, it is not necessary to convert all the silver halide to image silver by the action of light alone. It is only necessary for each grain to absorb enough photons to make a stable latent image speck somewhere on the grain. The active agent in the developer solution is a reducing agent which reacts preferentially with exposed grains to convert silver halide into image silver, so that grains having a latent image are reduced to silver before the unexposed grains. It is as if the latent image silver acts as a catalyst in causing the developer to convert more of the exposed grain to image silver.

The photometric uniformity of a processed emulsion depends predominantly on the method and degree of agitation of the solution during development, the promptness and effectiveness of agitation in the stop-bath, and to a lesser extent on techniques in subsequent stages of processing, such as adequate agitation of the solution during fixation and washing, and very careful control of the drying.

The efficiency of the development reaction increases with the level of agitation of the solution. It is relatively easy to process very small plates with high degrees of uniformity and efficiency. They may be developed very effectively by being shaken up with the developer in a closed thermos flask. Good uniformity can also be obtained by careful brush development for small plates. The plate is located horizontally in a tray containing developer, and agitation is provided by moving a brush (or possibly a wiper blade) back and forth very close to the emulsion, so that the developer is forced at speed past the emulsion surface, thus dispersing exhausted developer very rapidly. For best results the brush or blade must fit the solution tray well enough to prevent solution from flowing round the ends or over the back of the brush or blade and the brush must be wider than the plate. To achieve the optimum frequency of one stroke per second and to avoid damaging the emulsion, a mechanical guide is required for the brush. This is extremely difficult to provide for larger plates. Although excellent results can be achieved with this technique, it is only applicable to plates up to about 150 mm across, and can only be applied to one plate at a time.

Several plates may be developed simultaneously by suspending them vertically in a tank of developing solution, and agitating the solution by means of bursts of nitrogen bubbles released at the bottom of the tank. The plates are supported in a frame which allows all the plates to be moved into or out of the tanks simultaneously, and also maintains the plates at a fixed distance apart. With care, a bubble burst system is capable of producing a good uniformity for plates up to 250 mm × 200 mm in size. The optimum release frequency of the bubbles is of the order of a burst of 1 second duration at intervals of 10 seconds, and the pressure of the gas must be sufficient to mix the solution well at each burst, usually about 30 kilopascals (5 pounds per square inch). If the burst duration is too long, or the gas pressure too high, or the tank is too deep, the bubbles tend to collect and rise in columns rather than in a homogeneous mass, resulting in non-uniform development.

Many astronomers have used hand-rocked trays, or simple mechanical rockers to develop small or medium sized plates. In general the results obtained were neither repeatable nor uniform, and depended critically on the operator. An exhaustive study by Miller when the Mt Palomar Schmidt went into operation in 1948 led to the design of a mechanical rocking device capable of developing 360 mm square plates with a very high degree of uniformity, and which is still the best readily available method of obtaining uniformity of development on very large plates. The device is shown in Figure 3.9. The special features of the Miller rotating rocker machine are that the plate is held in a supporting frame which

Figure 3.9. The rotating rocker machine for the uniform development of large photographic plates. The plate to be developed is clipped on to an oversized handling frame and immersed in the deep tray containing the developer. The tray is automatically rotated and rocked simultaneously to provide a uniform and repeatable flow of developer over the plate. In this illustration the cover which normally protects the motor and gears has been removed.

locates it in the middle of an oversized tray of developer with a clearance of 50 mm between the edges of the plate and the tray walls. The tray then rotates and rocks simultaneously, so that the developer travels across the plate in a fast, laminar flow, and the turbulence as the solution rebounds from the tray wall is confined to the 50 mm space between the plate and the tray wall. The optimum rate of rocking decreases with increasing tray size, but must be carefully adjusted to avoid excess turbulence occurring over the emulsion. The rate at which the plate is rotated is not critical, but the angle of tilt from the horizontal and in particular the frequency of rocking are very critical. In correct adjustment the laminar flow of the developer over the plate is sufficiently strong to prevent the formation of a layer of partially exhausted developer in contact with the emulsion. The risk of evaporative cooling from the large surface area of agitated developer is overcome by covering the tray with a close-fitting lid during operation. Although designed to process a 360 mm square plate using a 460 mm square tray, this machine is also capable of processing 500 mm square plates very uniformly, or of developing simultaneously several smaller plates attached to an appropriate handling frame. Thus small test samples can be given exactly the same development as full-size plates.

The scientific value of an astronomical photograph increases very significantly with its age, and the photographic plate or film is potentially an archival information storage medium of infinite lifetime. Apart from the obvious risk of physical damage to a photographic plate, inadequate standards of processing and storage can cause slow deterioration of the photograph which may not be apparent until it is many years old. It then appears as an increasing degree of opacity or staining of the emulsion or perhaps as a gradual decay or fading of the image. It is now recognised that such effects are primarily due to inadequate fixing and washing of the developed photograph. In normal development of a photograph, only a fraction of the silver halide is reduced to silver and the remaining silver halide must be removed, without damage to the image silver. This is usually done by immersing the emulsion in a solution of sodium or ammonium thiosulphate (fixer) to dissolve the remaining silver halide, and then washing it to remove all traces of the solution. The reaction products produced by fixation in a fresh solution are all soluble compounds of silver, which are also capable of decomposing very slowly and after a period of years they may produce an opaque yellow stain in the emulsion if they are not completely removed during fixation and washing. The products of such decomposition are in general not soluble, and so cannot be removed simply by rewashing the plate. If an emulsion is left for too long in the fixing bath the silver of the image may be eroded by the thiosulphate, causing the image to fade. Low density images are particularly at risk. A fixing solution which is not very fresh and contains a significant buildup of silver may not be capable of dissolving all the residual silver halide from an emulsion.

A recommended method of reducing or avoiding all these problems by ensuring complete fixing in a reasonable time is to use two fixing baths in succession. The plate or film is immersed in the first bath and agitated until the emulsion clears – that is, the emulsion becomes transparent. At this point a large proportion of the silver halide has been dissolved. The plate is then transferred to the second, relatively fresh bath to complete the removal of silver compounds. The plate is immersed for the same time in each bath. In this way most of the work is done in the first bath, and the second one is fresh enough to complete the process. When the first bath approaches exhaustion, it is discarded and replaced by the second bath. The first bath is then cleaned and refilled with a fresh solution and becomes the second bath. The first bath should be discarded when the time required to clear a given emulsion in that solution exceeds twice the time required to clear the same emulsion in a fresh solution. After fixation the emulsion must be thoroughly washed to remove the residual chemicals from the fixing baths and any remaining dissolved silver compounds. If not removed from the emulsion, some of these products may eventually decay and cause discoloration of the processed plate. The surfaces of the emulsion and its support must be subjected to a brisk flow of clean filtered fresh water to ensure complete washing without requiring excessive wash times. The flow rate should be capable of replacing the volume of water in the washing tank at least once in five minutes, to remove residual chemicals as quickly as possible from the neighbourhood of the emulsion. Surprisingly, distilled or deionised water is too pure to wash out residual chemicals efficiently from the emulsion and should not be used as the main wash, although it is often used as a final wash to remove any deposits from the surface of the emulsion before it is dried. The speed and efficiency of washing may be improved considerably if the plate is immersed in a solution of hypo clearing agent for two minutes after removal from the second fix tank and before washing. This is a weak solution of sodium sulphite which aids the removal of chemicals from the emulsion by a rapid ion exchange process. Experiments have shown that an emulsion immersed in clearing agent and then washed in a brisk flow of fresh water for 5 minutes may be cleaner than a comparable emulsion not treated with clearing agent but washed for over 30 minutes. However, recent research into the keeping properties of the IIIa emulsions has raised several doubts about the chemical consequences of using clearing agents with these emulsions. It is now recommended that clearing agents should not be used with type IIIa emulsions, which should be washed for about 45 minutes after fixation. In regions where the supply of water is limited, a cascade washing system may be used. In this system clean water enters the first or highest tank level, and then cascades, in the manner of a multi-tier waterfall, through two or three consecutive baths. The plate just removed from the fixer is placed in the lowest wash tank, and proceeds up the cascade through each tank in turn, each tank being cleaner than the one below it, until the plate reaches the topmost tank and is finally washed in the cleanest water. The plate is then removed from the system and set to dry.

Experience with the United Kingdom and European Southern Observatory Schmidt telescopes shows that for critical work, plates should be dried at normal darkroom temperature, 20 °C, in a dust-free area such as a laminar flow clean air unit, with an ambient relative humidity of about 70 per cent. If a plate is set to dry whilst there are still some drops of wash water either on the glass or the emulsion side, the drops will evaporate causing localised cooling. The reduced drying rate in these areas may lead to local increases in the chemical fog level. Emulsions dried too quickly may become distorted, and changes of relative humidity can cause very small but significant variations in the background density of the photograph. Appendix A summarises the processing procedure adopted at the UK Schmidt telescope.

In January 1980 a new form of image decay was identified, to which the IIIa-J emulsion seems particularly susceptible, known as Gold Spot Disease or Gold Mould. It appears about 5 years after the plate has been processed. It takes the form of red or yellow spots around the images, and is most often seen near the edges of a plate or in heavily exposed regions. Since this disease includes some dissociation of the affected images, it is probably not possible to cure it and restore the photometric accuracy of the original photograph. Consequently prevention of the effect by adoption of safe processing and storage methods is very important. Tests have shown that the most likely causes of the problem include contamination of the emulsion by hydrogen peroxide (which is plentiful in foam plastics often used in packing) or by vapours from oil based paints and other oils such as cooking oils, or if the IIIa emulsion picks up iodide from a fixing bath previously used to fix a different type of emulsion, such as a IIa-O. The IIIa-J and IIIa-F emulsions are the only pure bromide emulsions used in astronomy, and consequently they may both be susceptible to gold spot disease. Appendix B gives the recommended procedures for processing these emulsions for maximum protection from gold spot formation. More detailed discussion of this problem will be found in the proceedings of the Nice conference, 1981, and in recent literature.

Plates and films processed for maximum archival permanence should be stored in a suitable environment. Paper envelopes, and the glues which hold them together, some paints, soft foam plastic, cardboard boxes, and wooden boxes or shelves are all capable of chemically contaminating emulsions; dust or excessive humidity may cause physical damage; high temperatures may cause the emulsion to soften, and low humidity may make the emulsion too dry and brittle. Ideally archival plates should be stored in envelopes made from an inert material such as Tyvek, supported against mechanical or pressure damage, and kept at 20 °C (± 5 °C) and at 40 (± 10) per cent relative humidity in an atmosphere free from dust and potential contaminants.

3.4 Image enhancement by copying

Successful production of a well exposed, optimally developed photograph is not necessarily the end of the useful photographic work to be done on it. It may still be worthwhile in certain circumstances to enhance the image, or part of the image, by photographic copying, rendering features more easily visible or measurable. It may also be desirable to make multiple copies of an original plate for wide distribution so that many people can make use of the photograph simultaneously.

The importance of the ability to copy original astronomical photographs will be illustrated by describing techniques used at the Royal Observatory, Edinburgh. This observatory houses a large library of original plates taken by the UK Schmidt telescope as well as a comprehensive collection of astronomical atlases, and these together form a unique reference collection for astronomers. Any one of these photographs may be made use of several times over, perhaps in the course of a number of different research programmes. Making copies of these plates for different users is an important function of the library, and a special photographic laboratory is maintained there for this purpose alone.

The simplest way to make a copy is to rephotograph the whole or, more usually, a part of the original with a copying camera. The method is flexible and a wide range of format, enlargement and contrast is possible. This sort of copy is more usually made for inspection or reference purposes, for example, to provide an illustration for publication, or to provide an observer with a convenient reference chart to take to the telescope. Such copies are not normally very suitable for quantitative work.

A more precise but more demanding technique is contact copying, in which the whole plate is reproduced whilst in direct contact with the copying medium. To ensure the sharpest possible reproduction, both original and copy emulsions must be in extremely close contact during the copying exposure, and this is usually achieved by use of a vacuum plateholder which exhausts the air between the two. Specially high standards of cleanliness must be maintained in the laboratory using this procedure, lest dust particles get trapped between the plates, damaging one or other of the emulsions. The original being a negative (black stars on a white sky), this copying procedure is usually carried out twice, via an intermediate positive, to produce a negative copy. The copies are of course all the same size as the original. A choice of copying media is available with different characteristics, and the final result may be on glass plate, transparent film or, less usually, paper.

The copying process maps a density D_0 on the original to a density D_2 on the final negative via the characteristic curves of the two copying emulsions as illustrated in Figure 3.10. By analogy with the gamma of the characteristic curve, the copying gamma is defined to be the gradient of the D_0 versus D_2 curve, dD_2/dD_0. By suitable choice of copying media, copying exposure times and processing, this gamma can be adjusted to any desired value over a limited range of densities. If the copying gamma could be made unity over the whole density range of the original, then the densities of copy and original would be the same everywhere, apart from a possible constant offset. However, since no media exist that can do this over the whole of the rather large dynamic range of astronomical emulsions, it is impossible exactly to reproduce an original in density and contrast.

There is a more fundamental way in which the copying process may affect the information transferred from the original. No matter how good the contact between original and copy, there is normally a degradation of the copied image in sharpness or resolution because light is scattered in the emulsions during the copying exposure, and because the grain pattern of the original plate is also mapped onto the copy, thereby adding to the noise on the copy. As a consequence, the signal-to-noise characteristics of the copy are likely to be worse than the original. This degradation may not be important if the copying procedure is properly carried out for the appropriate application. If the copy is to be measured in a microdensito-

Figure 3.10. Characteristic curves for each stage of a two-stage copying process.

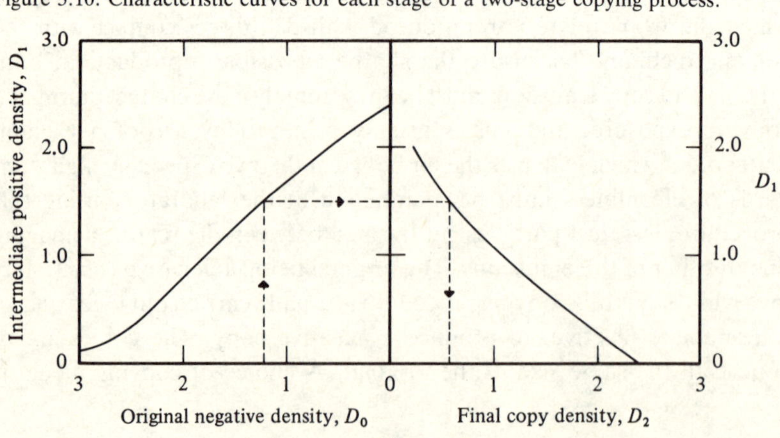

meter or similar measuring machine, then the signal-to-noise characteristics of the machine should be taken into account as well. Most measuring machines have poor signal-to-noise characteristics at high densities, and although the copying process may have degraded the plate from the original, the combined signal-to-noise characteristics of plate and measuring machine might actually be better for a good quality copy if reproduced at reduced density.

As an example of a particular copying problem, consider the task of making multiple copies of the European Southern Observatory/Science Research Council southern sky atlas for distribution to users around the world. The 606 original plates for the blue half of this survey were taken by the UK Schmidt telescope on IIIa-J emulsion. For the reasons discussed in the previous chapter, in order to maximise the signal-to-noise characteristics of this emulsion these plates were exposed until the sky background reached a density of at least 1.0. Since the brightest images are recorded at densities of about 4.5, the range of densities of interest is about 3.5. The uses to which the atlas copies are put are so varied that it was difficult to select the most important copying criteria. Since the main purpose is to produce a very deep sky survey, the faintest objects recorded, at densities close to the sky background, have to be reproduced as faithfully as possible. For this reason the copying gamma was kept close to unity at the sky density. The actual density to which the sky background is copied was chosen to be about 0.5. This is because the copies, which are contact printed as film transparencies, are intended for easy visual inspection, and the lower density is much more comfortable to the eye for use on normal light-tables. Since the copying emulsion saturates near a density of about 2.0, the copies have a density range of only about 0.5 to 2.0. This loss in dynamic range means an inevitable loss in contrast at the higher densities.

Two examples will now be given of copying techniques that have been employed to exploit the properties of modern fine-grain astronomical emulsions by enhancing features which, although present on the original photographs, are not obvious under normal inspection.

The very faintest images on long exposure plates, those at densities very close to the sky background, may be very difficult to detect because of their low contrast against the sky. This is especially true of diffuse, extended features in galaxies and nebulae. It may be possible to render these more easily detectable to the eye by copying at very high contrast. Naturally only a very limited density range can be reproduced in this way, and so the copy materials and processing methods are optimised to give the highest copying gamma at the sky density.

Figure 3.11 illustrates the dramatic results that can be achieved by high contrast copying of the IIIa type emulsions. Figure 3.11b was produced as follows. The faintest images in a fine-grain negative are in fact situated mainly near the top surface of the gelatin and do not extend deep into it. However, the fog grains, which merely contribute to the noise, are distributed throughout the gelatin layer. By using a contact copier with a diffuse light source, it is possible to copy the surface grains onto high contrast film. The fog grains deeper in the layer are not sharply recorded and do not contribute to the image as they would do if the illumination were a parallel beam.

Originals must be of the highest quality for successful application of this method. All the faint handling and processing defects of the original are enhanced along with the real images. In cases where there is any doubt about the reality of features that are revealed by this process, a second plate must be obtained and processed the same way for confirmation.

Turning now to the brightest galaxies and nebulae, these may be recorded at densities as high as 4 or 5 on IIIa emulsions, and using normal light sources such images are impenetrable to visual inspection. Detail in extended areas of bright nebulosity is invisible. Simply copying the whole photograph to a lower density is one possible solution, but in conventional copying the information recorded at lower densities would be lost. A technique called unsharp masking can be used instead, which is a photographic equivalent of a spatial frequency filter.

Photographic masking is widely used in the graphic arts industries to control contrast. First a low contrast positive transparency is made on film from an original negative by contact copying. This positive is used as a mask. Original and positive are held in register with each other for a second contact copy to be made. In this way the effective total density range of the original is reduced. If, in addition, the mask is made slightly unsharp, registration is less critical and the mask acts only on the coarse detail of the subject. The mask thus acts as a low frequency spatial filter whose effect is to enhance rendition of fine detail whilst reducing gross density variations. The unsharpness of the mask can be produced by spacing the film from the original as the positive is being made, for example by exposing it in contact with the back of the glass instead of the emulsion and using a diffuse light source. The length scale of the spatial filter is then set by the thickness of the plate.

Figure 3.11. An illustration of high contrast copying. Print (a) is a normal contrast reproduction of a IIIa-F plate taken on the Anglo-Australian 3.9 metre telescope, showing part of the Puppis supernova remnant. Print (b) is a high contrast copy from the same plate, of exactly the same area, produced by the method described in the text. The high contrast copy has dramatically revealed the presence of intricate structure in the remnant which, although faintly recorded on the original photograph, is difficult or impossible to see at normal contrast. Notice how the technique also unavoidably enhances the halation ring around the bright star. The photographs were produced by David Malin of the Anglo-Australian Observatory.

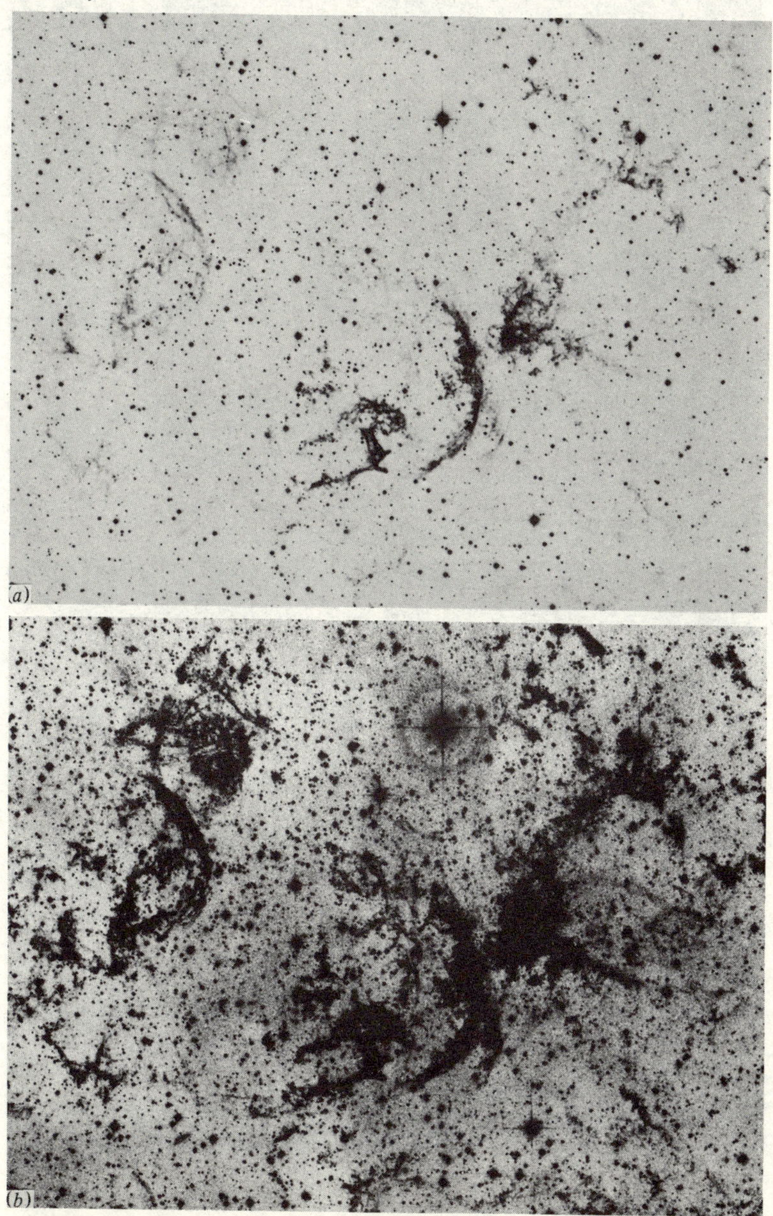

Figure 3.12. The unsharp masking technique. Print (a) is a conventional copy of a photograph of NGC 7293, the Helix nebula, taken from a IIIa-F plate obtained on the Anglo-Australian telescope. Print (b) is a copy of the same original plate through an unsharp mask. The technique has highlighted the fine detail concealed in the heavily exposed regions of the nebula. The prints were produced by David Malin of the Anglo-Australian Observatory.

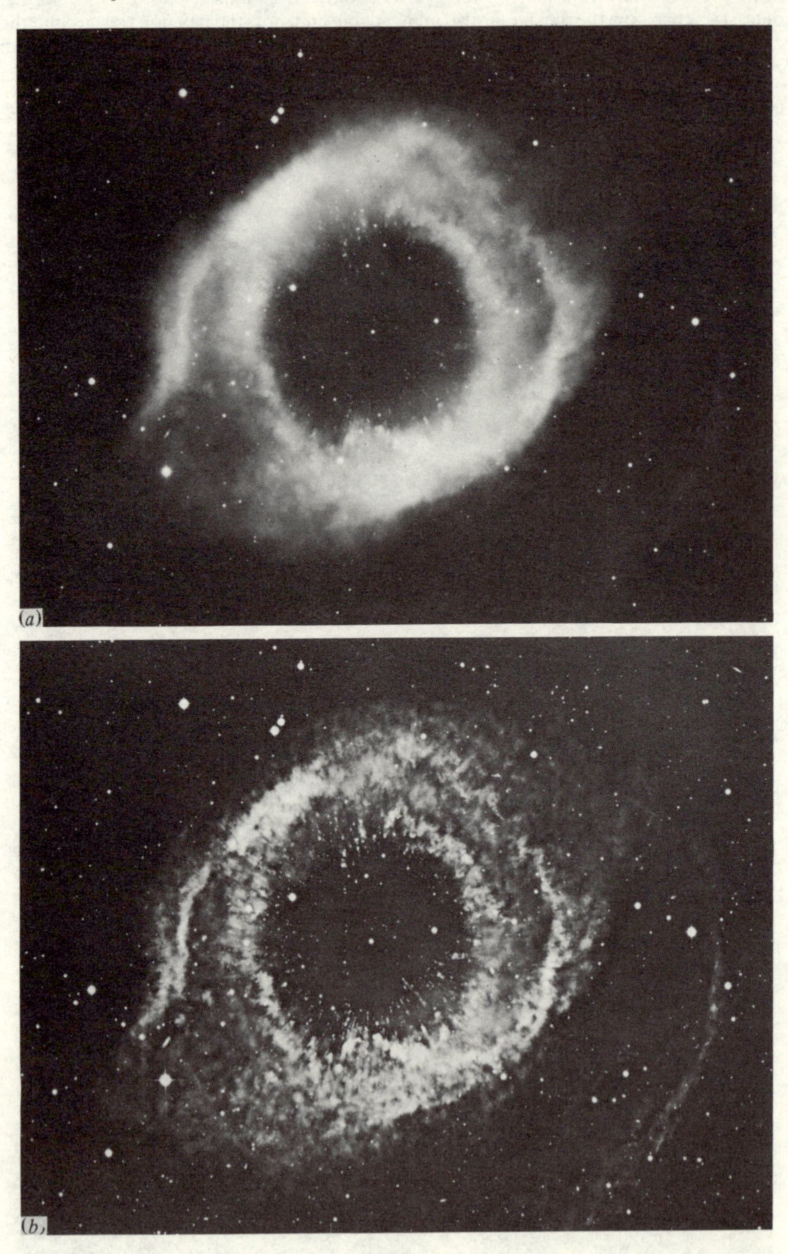

Figure 3.12 shows an example of the results of this method. The masking has emphasised the detail in the brighter parts of the nebula in a way not possible with conventional techniques. Faint background galaxies are still easily visible on the masked version and even the images of bright stars are unaffected by the mask, since their diameters are less than the spatial filter length. Because of the filtering process, quantitative information is lost in these photographs, and the results are primarily for visual inspection.

References for Chapter 3

American Astronomical Society Photobulletin, 1970 onwards.

Heudier, J.L. & Sim, M.E. (eds.), 1981. *Astronomical Photography 1981* – Proceedings of meeting of International Astronomical Union Working Group on Photographic Problems held in Nice, April 1981.

Smith, A.G. & Hoag, A.A., 1979. Advances in photography at low light levels, in *Annual Review of Astronomy and Astrophysics*, Volume 17, p.43.

Appendix A. Summary of plate processing procedures used at the UK Schmidt telescope, except for IIIa-J and IIIa-F emulsions

1. *Develop.* Individual plates are developed in a rotating rocker which is controlled by a foot switch and timer. Original spectroscopic plates are developed in D19 for 5 minutes at 20 ± 0.2 °C. Three litres of developer are required to cover the plate adequately, and are enough to develop six 356 mm × 356 mm plates in succession.
2. *Stop.* The plate (still attached to a handling frame) is plunged into a horizontal bath of Kodak Indicator Stop Bath, made up to half the recommended strength (1 per cent acetic acid) and agitated very vigorously for at least 30 seconds. This solution is freshly prepared before each processing session, and discarded at the end of the session.
3. *Fix.* The plate and handling frame are immersed for 5 minutes in each of two successive baths of normal fixer solution or for 4 minutes in a single bath of rapid fix, and agitated with nitrogen bubble bursts.
4. *Rinse.* Surplus fixer is removed by immersing the plate in a vertical tank of fresh water for about 30 seconds.
5. *Clear.* Immersion of the plate for 2 minutes in a vertical tank of Kodak Hypo Clearing Agent solution helps to remove byproducts and improves the efficiency of washing.

6. *Wash.* The plate is immersed for 10 minutes in each of two successive vertical tanks, each tank being supplied with a brisk flow of filtered tap water.
7. *Photoflo.* The plate is rinsed for 30 seconds in a standard strength solution. If excess suds are to be avoided, the solution strength must not exceed the recommended concentration.
8. *Swab.* With the plate immersed in a horizontal tray of distilled water, the surface of the emulsion is carefully swabbed with a lint-free pad to remove any particles which may be adhering to the emulsion.
9. *Drain.* The plate, still in its handling frame, is supported in a vertical draining rack and allowed to drain to the bottom corner until surplus water stops dripping from the plate. The back of the plate is carefully wiped with a lint-free pad to remove water droplets.
10. *Dry.* A horizontal flow clean air cabinet, filtered to 0.3 μm, and maintained at 20 °C and 70 per cent relative humidity, is fitted with horizontal drying racks. The side of the plate which was uppermost in the draining rack enters the drying cabinet first, so that the plate is dried horizontally and the direction of drying in the cabinet is the same as the direction of draining. Alternatively, the plate may be dried vertically in the same clean air cabinet, provided that the orientation of the plate is the same as it was in stage 9, and that the plate is parallel to the direction of air flow.

Appendix B. Recommended processing procedures for IIIa-J and IIIa-F emulsions to avoid gold spot formation

To avoid any pick-up of contaminants from other emulsions via the processing solutions, a separate processing line should ideally be established for these emulsions. In practice it is probably acceptable to ensure that there is a dedicated fix bath, used only for type IIIa emulsions, to provide an additional bath of rapid selenium toner, and to ensure that IIIa emulsions are always processed before any others, so that they always enter fresh, uncontaminated solutions and wash tanks.

1. *Develop.* The procedure is the same as that described in Appendix A.
2. *Stop.* The procedure is the same as that described in Appendix A.
3. *Fix.* The procedure is the same as that described in Appendix A, except that a single dedicated tank of fresh rapid fix is preferred.
4. *Wash* (1). The plate is washed in a vertical wash tank in a brisk flow of filtered tap water for 20 minutes.

5. *Rapid selenium toner.* The plate is immersed for three minutes in a fresh solution of rapid selenium toner, diluted 1 + 19. This appears to give some protection against post-processing contaminants.
6. *Wash* (2). Again the plate is washed in filtered tap water, for 25 minutes.
7 to 10. The final stages of Photoflo, swab, drain, and dry are as indicated in Appendix A.

4

Image tubes and electronography

The previous two chapters have been concerned with the use of the photographic film or plate as a two-dimensional low light level detector. For certain applications, principally the study of very faint objects, it has been found advantageous to amplify the incident photons in an image intensifier before photographic detection. One variation of this technique – in which a film that is sensitive to high-energy electrons detects an amplified electronic analogue of the optical image – is known as electronography and forms the principal subject of this chapter.

4.1 Image intensifiers

The most important part of the image intensifier is a chemically coated glass plate called the photocathode, which emits electrons when exposed to incident photons. The basic process of image intensification, or optical amplification as it is sometimes called, starts with the production of such electrons from light incident on a photocathode and proceeds with the acceleration of these electrons to a high energy. The electrons finally impinge upon a target, generally a phosphor screen, which emits photons. The electron acceleration ensures that many secondary photons are emitted by the target for each primary photon incident on the photocathode.

A detailed discussion of the physics and operating parameters of the more common types of photocathode is deferred to Chapter 6, but it is sufficient here to say that the majority of photocathodes operate most efficiently towards the blue end of the optical spectrum and have quantum efficiencies of some 10 to 20 per cent. The image intensifier usually consists of an evacuated cylinder with a photocathode at one end of the cylinder, some form of target at the opposite end and a potential difference of typically 20 kV maintained between the two ends. The electrons emitted by the photocathode are focused by an electrostatic or magnetic system to form an electronic analogue of the optical image on the target. The simplest of such image tubes employs a phosphor screen as the target so that the amplified image can be either viewed directly by eye or else recorded photographically. Further amplification can be achieved by cascading several tubes. It is quite common to find three similar tubes cascaded which, since a single image tube has a typical gain of 40, gives an overall optical gain of over 6000. The main practical difficulty in this case is in maintaining a potential difference of perhaps 50 kV across the cascaded combination under various ambient atmospheric conditions. The principal advantages of using an image tube preceding a photographic plate are the higher quantum efficiency of the photocathode and the increased light incident on the photographic plate, which may enable the use of the more linear part of the photographic response curve. The principal disadvantage is the decrease of signal-to-noise ratio because of the statistical nature of the electron amplification mechanism. For this reason phosphor image tubes are mainly used for the detection of faint objects rather than the accurate measurement of their brightness. The fundamental spectral response of an image tube–detector combination is determined by the photocathode and not the detector, and this enables detectors to be used outside their normal spectral range. This property, known as image conversion, is particularly useful in infrared surveillance work and is the reason for much of the commercial and military interest in image tubes.

The overall layout of a typical electrostatically focused image tube is shown in Figure 4.1. The incident image is focused onto a glass screen and transferred to the photocathode via a fibre optic channel plate. The electrostatic focusing electrodes are such that both the object and image planes are curved and the image is reduced in size. The image is formed

Figure 4.1. A single-stage electrostatically focused image intensifier.

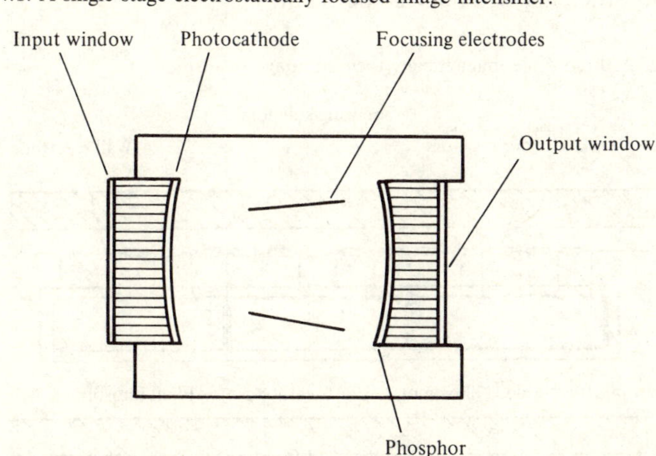

on a curved phosphor screen and transferred to a flat face plate in contact with a photographic plate. Before fibre optics were available, the transfers between the image tube and the flat fields of the optical image and detector were carried out by means of transfer lenses, and this put severe limitations on the overall efficiency of the tube–detector combination since the lenses collected only about 3 per cent of the light. Fibre optics have greatly improved the optical transfer efficiency but have introduced new problems including blemishes, dislocations and 'chicken wire' patterns, usually the result of poor manufacture. In general electrostatically focused tubes have rather poor image quality, although they are rugged and relatively straightforward to set up and use.

Most of the work on image tube development has been for military applications, where strength and simplicity of operation have taken precedence over image quality and hence tended to favour electrostatic tubes. However, for astronomical purposes, the magnetically focused system is preferred because of its higher image quality. Figure 4.2 shows a typical layout for three such tubes operated in cascade. In this case the object and image fields are flat and there is no need for optical image transfer. The solenoid is wound so as to produce as nearly uniform an axial magnetic field as possible. Light focused on the first photocathode generates photoelectrons which are focused onto the first phosphor; this stimulates the second photocathode and so on to the flat glass output screen. The detector, typically a photographic film, may be pressed against this screen, or an optical system may focus the image elsewhere. This arrangement allows for a large field, typically 90 mm, over which the image quality is comparable with a direct photograph, but with an overall quantum efficiency of about 20 per cent.

Figure 4.2. A three-stage magnetically focused image intensifier.

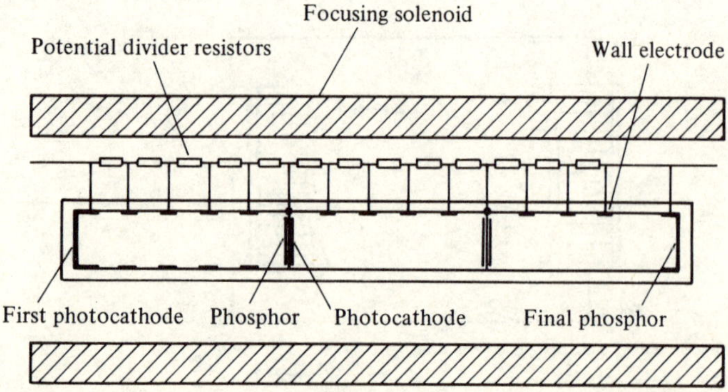

The problems of poor image quality, small or blemished field, geometrical distortion, and ultimately non-linear response have limited the use of phosphor image tubes in astronomy. There are nonetheless a number of areas where they have proved of considerable value. The visual detection of faint images is frequently assisted by an electrostatically focused tube. Although the gain over the unaided eye is only marginal in blue light, gains of 1 or 2 stellar magnitudes can be obtained for red images. An instrument of this sort is often positioned to facilitate the centering of a faint image in the aperture of a photometer or the slit of a spectrograph. Phosphor image tubes, either electrostatically or electromagnetically focused, are also frequently used as detectors where photometric measurements are not required, such as for the measurement of the position of spectral lines, or the detection and counting of faint objects. With the best electromagnetically focused tubes a relatively large format and uniform response can be obtained, and photometry of stars and galaxies becomes possible with careful calibration. Finally, a number of compound detectors incorporate image intensification stages. Examples of these will be given later, in Chapter 7.

4.2 Electronographic cameras

As long ago as 1936, A. Lallemand put forward a new solution to the problem of recording the electronic image produced by an image tube. He proposed to record the electronic image directly on to electron sensitive film and it is this process which is known as electronography. The most important property of an electronographic image is that the photographic density recorded on the film is closely proportional to the flux incident on the photocathode. The main problem to be overcome in designing an electronographic camera is to position a piece of film so that the electronic image may be focused on it, bearing in mind that the interior of the tube must be kept a hard vacuum to preserve the photocathode.

Lallemand's solution was to evacuate the tube and deposit the photocathode with a magazine of eight films already in position; in retrieving the films, the photocathode was destroyed. Apart from the obvious inconvenience of this method there was a problem of gas from the film contaminating the photocathode, which was alleviated in part by operating at very low temperatures. Later, G.E. Kron adopted a somewhat similar solution but contrived to maintain the photocathode indefinitely by the use of two chambers separated by a coin valve, the film being introduced into the first chamber with the valve shut to maintain a hard vacuum around the photocathode. In order to expose, the loading chamber was evacuated and the valve opened. This camera was complicated and time-consuming to use and still suffered from most of the drawbacks of the Lallemand camera, except that the photocathode was preserved.

An important breakthrough in the design of electronographic cameras was achieved by Professor J.D. McGee and his colleagues in the early 1960s. They inserted a 4 μm thin electron-permeable mica window in the end of the tube opposite the photocathode. The film was then pressed gently against this window from the outside and the transmitted electronic image recorded. This camera was known as the Spectracon, and incorporated a number of other improvements including electromagnetic focusing in place of the electrostatic system of Kron and Lallemand. The Spectracon made electronographic observations quick and easy to make, but the uses to which it could be put were severely limited by the size of its field of view. The mica window had to withstand atmospheric pressure, and this effectively limited its size to 5 mm $\times$ 25 mm. This was inconveniently small for most direct work but resulted in the use of the tube as a spectrographic detector. Other troubles with the Spectracon included excessive dark current and insulation breakdowns in the 40 kV high tension (HT) system, although eventually most of the problems were overcome.

The Spectracon has now to a large extent been superseded by a large field camera developed by D. McMullan at the Royal Greenwich Observatory. Like the Spectracon, this camera basically consists of an image tube, but with a 40 mm circular mica window, protected from atmospheric pressure by a gate valve which allows a pneumatically operated applicator to bring the film, mounted on a plastic ring, into contact with the mica. Control circuits and relays are mounted on either side of the gate valve, while power supplies, vacuum pump, etc. are contained in a free standing console. The general arrangement is shown in Figure 4.3. The vacuum tube is of fused silica and the photocathode is formed directly on the faceplate at one end. The electrodes form a separate unit and are titanium annuli separated by glass cylinders fused together with solder glass. Resistors are mounted between the electrodes, and the glass cylinders, being slightly conducting, form a uniform potential gradient down the tube. This whole assembly is mounted inside the silica vacuum tube and down one side is a small glass tube carrying the 40 kV HT cable to the photocathode. The 40 mm mica window is mounted on a titanium ring at the end of the vacuum lock and is coated with a black aluminium film to

Figure 4.3. The McMullan 40 mm electronographic image tube.

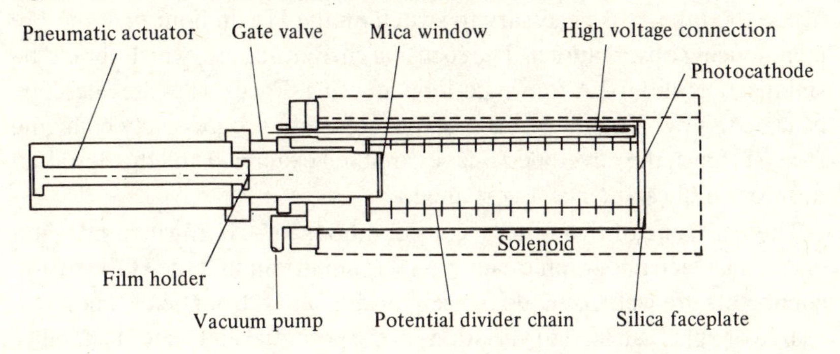

cut down internal light reflections. There is an ion appendage pump connected to the tube to prevent a buildup of gas diffusing through the silica envelope. The tube is contained in an electromagnetic focusing solenoid wound on a cooling jacket (which is in fact not normally used).

Film is mounted on a bayonet type applicator which is inserted behind the gate valve leading to the vacuum chamber and mica window. In order to open the gate valve, the area around the film is evacuated by a vacuum pump in the console. The gate valve then opens and the film applicator moves forward pneumatically to the mica window. The entire sequence is automatic, and the total time taken to unload, change film and reload is 2 – 3 minutes. Higher pressure is allowed behind the film to hold it in place against the window. In practice this particular hold-down procedure has never allowed adequately intimate contact between film and mica, and two other systems have been tried. A method in which the film was held electrostatically to the mica showed great promise but the problem of eliminating discharges on separation, which affected the film, was never satisfactorily solved. An alternative procedure in which the film is pressed to the mica with a sponge, as in the Spectracon, has proved more successful and may become the standard procedure in future.

In operation, an image is focused on the photocathode, and the resulting electrons are accelerated down the tube to the mica window by a potential of −40 kV, being brought to a focus there by the focusing solenoid. The focus current is adjusted so that electrons make either one or two loops during their travel. Two-loop focusing requires a higher focus current and gives slightly better resolution but tends to produce excessive heat. On reaching the mica window, 75 per cent of the electrons pass through with an average residual energy of 15 keV, which is sufficient for each electron to generate several developable grains in the nuclear emulsion. In order to let the charge pattern reach equilibrium inside the tube, it is necessary to switch on the HT an hour or so before commencing observations. The focusing current supply, which should be stabilised, is about 2 A for single-loop focusing. Focusing at the telescope is achieved by taking a series of exposures with progressively changing focus settings; the developed images are then examined and the optimum position in the sequence ascertained.

There are one or two general considerations to be borne in mind when operating electronographic cameras. It is important that the HT lead and connectors are well insulated, since sparking, apart from being a possible source of light, can lead to variations in the potential and hence instability of the image. Care must be taken to protect the focusing coils from stray magnetic fields for, although they are shielded by Mumetal screens, this is only proof against weak fields. The most vulnerable part of the system is the 4 μm thick mica window, and great care is necessary in the preparation of the films. These are punched out in circles and stuck on to plastic rings prior to the night's observations. It is important to check for protruding particles and to keep the film dry, because if it is even slightly moist it can stick to the window, removing a layer of mica or causing it to break on retraction.

4.3 Recording emulsions for electronography

The response of a silver halide emulsion to high energy radiation, such as the electrons from an image tube, differs in one major respect from the response to light. Whilst several light photons are required to generate enough photoelectrons to make a single silver halide grain developable, a single energetic electron is capable of making several grains developable. This has several important consequences, not least of which is the absence of reciprocity failure in exposures made to such sources. In the lower part of the D versus log E curve, the relationship between density and exposure remains approximately linear for exposures to energetic electrons, instead of showing the toe which occurs for exposures to visible radiation. It is easy to deduce that the gamma of the D versus log E curve at lower densities is approximately related to the density by gamma = $2.3D$. At high densities the emulsion will saturate, whether the exposure is to light or to other radiation.

The silver atoms formed by exposure to visible light lie mainly in the surface layers of the emulsion, but ionising radiation generates silver atoms at greater depths inside the emulsion thickness as well as at the surface. Nuclear track emulsions have been specially made to record this type of image, although other emulsions, including the spectroscopic emulsions, will also record such radiation quite effectively. The granularity of a given emulsion is higher for an exposure to electrons than for an exposure to photons.

There are several types of emulsions used to record electronographic images. The spectral sensitivity of the recording emulsion to visible light is of course not significant when it is being used to record electrons, since the spectral response of the image tube system is associated with the photocathode. Of the spectroscopic emulsions listed in Table 3.1, types IIa-O, IIa-D and IIIa-J have been used quite successfully for electronographic recording. Kodak also make a series of emulsions especially for electronography. These include a high speed electronographic emulsion, similar to the nuclear track emulsions, and a fine grain emulsion. The grain sizes of these emulsions are 0.34 μm and 0.15 μm respectively, and each emulsion is 10 μm thick. The physical characteristics of image tubes differ, so these emulsions are available on 65 μm or 180 μm Estar film base, or on 1.5 mm glass plates. Ilford produce a series of nuclear track emulsions which are widely used in electronography. These include types G, K and L, with mean grain sizes of 0.27 μm, 0.20 μm and 0.14 μm respectively. Sensitivity is denoted by a number, such that G5, K5 and L4 are the most sensitive and K0 the least sensitive. The types most commonly used in astronomy are G5 and L4. The emulsion thickness is usually 10 μm on a 50 μm thick backing, although for a time Spectracon users preferred a 25 μm backing as this was considered less likely to damage the mica window. An emulsion is supplied on glass plate, from which it is stripped before being mounted against the output window. Emulsions of this type have a thin backing, such as Melinex, to permit stripping without distorting the emulsion. Fortunately nuclear track emulsions are relatively insensitive to visible light and can therefore be handled under orange safelights, very much simplifying what would otherwise be an awkward task.

After exposure the emulsion must be developed in order to render the recorded image visible. The development procedure is essentially the same as for normal emulsions, but with one additional stage. Some of the nuclear track emulsions are made with a relatively small amount of gelatin separating the silver halide grains, and during exposure an emulsion may have been subjected to a high vacuum. In either case, the grains are likely to be in rather close proximity, and there is some danger of infectious development, that is, unwanted development which spreads by contact from grain to grain. The emulsion is therefore soaked in distilled water to swell the emulsion and separate the grains before development. The processing procedures may therefore be summarised as follows:

Pre-soak in distilled water at 18–21 °C for 4 minutes.
Develop in D19 or Phen-X at 20 °C for 5 minutes with continuous agitation.
Rinse in stop bath for 30 seconds with agitation.
Fix in normal fixer for 10–15 minutes, or in Rapid Fix for 3–5 minutes, with frequent agitation, at 18–21 °C.
Wash in running water at 18–21 °C for 20–30 minutes with moderate agitation.
Rinse in wetting agent solution (Photoflo) for 30 seconds.
Dry in a dust-free place at room temperature.

In common with all other photographic procedures required to give results of high quality, considerable care is needed at all stages to ensure cleanliness and uniformity.

Intrinsic variations occur in the sensitivity of nuclear track emulsion up to 5 per cent over a few centimetres. This problem may be compounded by development non-uniformities, and it is difficult to distinguish the two effects. In neither case is it possible to calibrate them out. This means that isophotometry of a galaxy occupying a small area of film can be done with greater accuracy than the measurement of a sequence of standard stars which typically covers several centimetres.

The remainder of this section will deal with some performance parameters of L4 and G5 emulsions and, where applicable, a comparison will be made with Kodak IIa-O photographic emulsion. The noise power

Figure 4.4. The noise power spectrum for L4 emulsion.

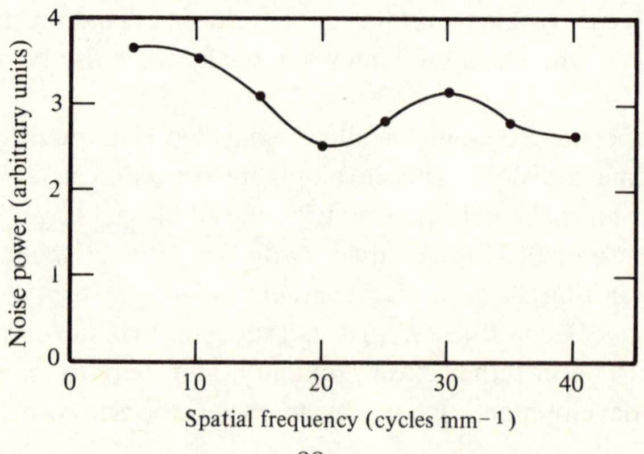

spectrum is of importance in assessing the performance of a detector system; it is also needed for various filtering techniques of data reduction. Figure 4.4 shows the noise power spectrum for L4 emulsion. It can be seen that the curve is nearly flat, a spectrum of this type being known as white noise. G5 nuclear and Kodak IIa-O photographic emulsions also have flat noise power spectra, but although L4 has the lowest noise power, G5 has a higher noise power than IIa-O despite its small grain size. This is attributed to a difference in the distribution of the grain sizes sensitive to the electronographic and optical processes. Defining 'information rate' as the information content at a given density divided by the time taken to reach that density, L4 is about five times as fast as IIa-O and G5 about five times as fast as L4. This figure is roughly confirmed in practice where it is found that about three times the exposure is required to obtain a given photometric accuracy with a IIa-O photographic plate as with an L4 electronograph.

The choice between G5 and L4 depends upon the detailed requirements of the observation, but G5 has a non-linear response at high densities which makes it unsuitable for many projects in spite of its gain in speed. Although the information rate for G5 is higher than for L4, the information content of L4 is greater, which makes it a natural choice where exposure times are not important. In addition, G5 tends to be more difficult to use due to its greater sensitivity to pressure and the consequent danger of marking the film during preparation. Finally, it should be mentioned that with the advent of modern fine-grain photographic emulsions such as Kodak IIIa-J, the information content of photographs has been increased enormously. In this sense the performances of photography and electronography are to some extent converging.

4.4 Performance characteristics

Perhaps the most important property of the electronographic process is the closely linear relation between the incident photon flux and the photographic density recorded on the emulsion. The essential characteristic of this process is that only one electron is necessary to create one or more developable grains, in contrast to the photographic process. In fact the number of developable grains is strictly proportional to the electron flux, which is in turn proportional to the photon flux through the properties of the photocathode.

The empirical relation between photon flux and photographic density in electronography has been the subject of extensive investigation. A number of tests have shown that for L4 emulsion a linear relation holds to better than 1 per cent up to a density in excess of 5, and for G5 to a density of about 2–3. (Electronographs are usually measured on microdensitometers, which measure specular density. The densities quoted in this chapter are specular, and can be converted approximately to diffuse densities by multiplying by 0.76.) To ensure this performance it is important that the emulsion be made strictly to specification, as sub-standard batches, which are not infrequently encountered, saturate at very much lower densities. It is important that all emulsion batches are tested for linearity before use, and preferably also on the telescope against a standard photometric sequence.

The spatial resolution of electronographic cameras is determined by a number of factors, in particular the mica window, the focusing system and the emulsion. The lateral spread of electrons passing through the 4 μm mica window limits the resolution to a maximum of 150 line pairs per millimetre (lp mm^{-1}). In order to obtain this figure, very uniform magnetic fields ($\pm$ 1 per cent) must be achieved with the focusing solenoid. The number of turns along the solenoid must be carefully adjusted, and cameras must be protected from the Earth's magnetic field and from stray fields from electronic equipment by two high permeability Mumetal screens. Different emulsions have different resolutions. L4 and G5 nuclear emulsions have a similar resolution at 50 per cent MTF where they have a resolution of 50 lp mm^{-1}, compared to 35 lp mm^{-1} for IIIa-J and 25 lp mm^{-1} for IIa-O. Limiting resolutions (at 5 per cent MTF) are 120 lp mm^{-1} for L4, 90 lp mm^{-1} for G5 and 70 lp mm^{-1} for Kodak IIa-O. Providing the contact between film and mica window is sufficiently intimate, the limiting factor for resolution is the lateral spread of electrons in the emulsion, the typical grain size being 0.2 μm. However, a figure of 50 lp mm^{-1} is the best that can be obtained using the pressure hold-down of most McMullan cameras because of electron scattering in the mica/emulsion gap. The electrostatic hold-down technique has improved the resolution to 120 lp mm^{-1} but as mentioned above there are problems of fogging with this method due to discharges after separation of film and mica, and this method is not generally available on cameras currently in use.

One of the more difficult problems with electronographic cameras, especially in early work on their development, was to obtain good resolution and image geometry over the whole field, because this depends on the complex electron focusing system. The Lallemand and Kron cameras used electrostatic systems, curved photocathodes and images, and a very accurately made electron lens. A resolution of 80 lp mm^{-1} could be obtained at the centre of the field, although this fell off towards the edges. In addition, there was a 1 : 2 demagnification of the final image. Initial problems with image geometry were also experienced with the Spectracon, but the use of parallel electric and magnetic fields has proved to be an inherently better system and after a lengthy development period adequate image geometry has been obtained for almost all astronomical purposes. In the McMullan camera geometrical distortions such as pincushion distortion do not exceed 25 μm, but there is a demagnification of about 10 per cent due to the use of a non-uniformly wound solenoid.

A certain unavoidable background noise current arises within the tube from two sources: thermal emission of electrons from the photocathode, and emission of electrons from other components of the tube. These may be detected at the output window but sometimes strike the photocathode, causing secondary emission. The latter effect, when detected at the output window, is called an ion spot. An S20 photocathode at 10 °C, roughly the normal operating temperature, produces about 0.7 noise electrons mm^{-2} s^{-1} which corresponds to a photographic density of about 0.0005 per hour for L4 and 0.0025 per hour for G5. At 25 °C the flux is ten times as great. In almost all practical circumstances these figures are negligible compared with the sky background. The background noise current presented one of the most serious difficulties in the use of the Spectracon, but great improvements have been made with the McMullan camera. Breakdown of the insulation inside the tube, resulting in small sparks, produces a current, sometimes uniform, sometimes making a pattern, which is sufficient to ruin an observation. When these effects arise in a particular tube they are usually incurable, although sometimes they are intermittent and seem to be caused by temporary charge buildup. In any case, such tubes are better replaced, as a large proportion of tubes give no trouble in this respect. Recent research has shown that the problem can be completely cured by painting the inside of the silica envelope and electrode stack with chromic oxide, a conducting paint.

The sensitivity of an electronographic camera is essentially a function of the photocathode, and at 410 nm a quantum efficiency of 17 per cent for an S11 and 29 per cent for an S20 can be obtained, approximately the peak sensitivity in both cases. To date, these two photocathodes are the ones most commonly found in ground based electronographic cameras, due to their high sensitivity and good spectral response. The S1, with a red response out to 1.2 μm and peak quantum efficiency of about 0.5 per cent, is also attractive but is not widely available in cameras currently in use. Ultraviolet sensitive photocathodes are also used in cameras in rocket and balloon-borne experiments.

4.5 Measurement and calibration

Electronographs normally convey little to the eye and are particularly deceptive to observers used to examining photographs. Electronographic images appear soft and out of focus, while features which are obvious on photographs are hard to distinguish or not apparent at all to visual inspection. There are a number of reasons for the differences in appearance. Perhaps the most important is that the density of an electronograph, in contrast to a photograph, is a direct linear record of incident flux. Thus star images, for example, appear as Gaussian density profiles. This is illustrated by Figure 4.5 which shows part of an electronographic

Figure 4.5. Part of an electronographic image of the globular cluster NGC 288, taken with the 80 mm McMullan camera mounted on the Danish 1.5 m telescope at the European Southern Observatory, La Silla, Chile. The exposure time was 60 minutes in the *V* band in 1 arcsecond seeing conditions. The white blemishes visible in this reproduction are photocathode defects.

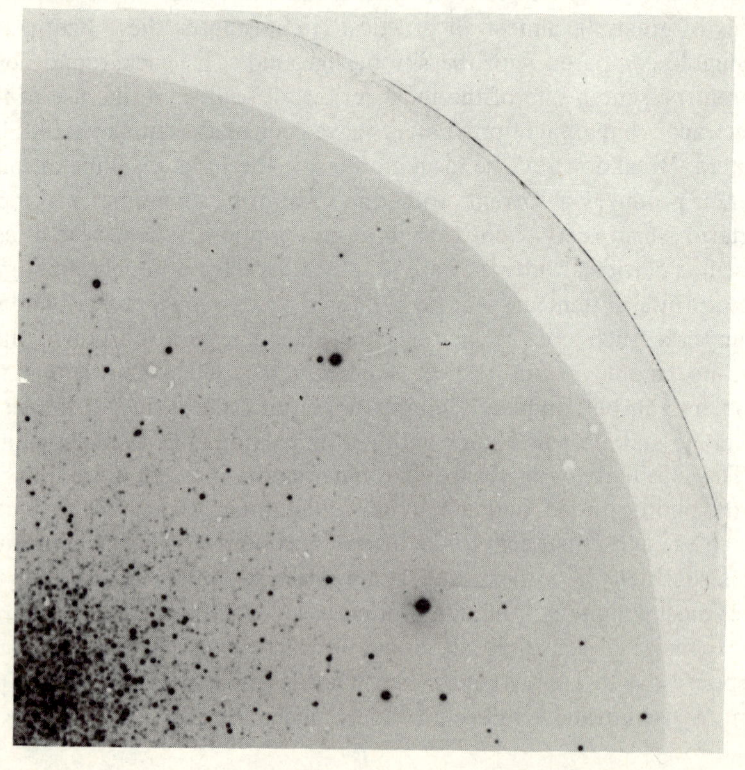

image of the globular cluster NGC 288, taken with a McMullan camera on the Danish 1.5 m telescope at La Silla, Chile under excellent seeing conditions. The original has been much enlarged and shows a number of defects typical of an electronographic exposure.

Electronographs are invariably measured by some form of microphotometer to extract astronomical information. The measurement is made indirectly by means of a beam of light shone through the film. The transmitted light may be measured directly by a photomultiplier; alternatively the light beam may be split so that half passes through the film and the other half passes through a wedge of varying and calibrated optical density. In the latter case, the two transmitted beams are compared and a servo mechanism moves the wedge until the two transmitted fluxes are the same. Either method may be used to take large numbers of readings through an aperture a few tens of microns across, and thus produce a rectangular array of density measures. In fact it is usual for the light beam to remain stationary and for the film to be mounted on a carriage which can be moved automatically in two dimensions with an accuracy of 1–2 μm. The carriage then moves in a raster pattern, stepping by predetermined sampling intervals.

A number of general considerations arise with the use of microphotometers. It was mentioned in the previous section that the density–exposure relation for an electronographic camera with L4 emulsion is linear up to a specular density of about 5; unfortunately, a major problem arises in measuring densities of this magnitude. Microphotometers effectively measure transmission and so at a specular density of 5 they are measuring only one part in 10^5 of the incident light. With a sufficiently large aperture this is possible under advantageous conditions, but for photometric work, where a small aperture is necessary together with a fast data rate, it is not normally possible to measure specular densities greater than 3–4 with acceptable accuracy.

The choice of aperture size involves a number of conflicting considerations. Clearly the aperture must be smaller than the required resolution and this normally means smaller than the seeing size. Another reason for using the smallest possible aperture is associated with the fact that the transmitted light is averaged over the aperture. When the averaging is done over a sharp density gradient the result obtained can be significantly different from that obtained by averaging the density, since the measured density varies as the logarithm of the transmission. For a Gaussian image with a peak central specular density of 3, an aperture of one-sixth the half power diameter of the image will result in an error of 2 per cent in the integrated density. The error arises in the measurement of the high density gradient part of the image, which will be measured too low. On the other hand, there is a practical limit to the smallness of aperture, for errors due to the positional accuracy of the carriage, optical alignment and so on, start to become important with an aperture smaller than about 10 μm. This limitation combined with that mentioned in the previous paragraph imply that to enable accurate measurement of images, the plate scale should be more than about 60 μm per arcsec, which on any but the largest telescope involves using the Cassegrain focus. It may be remarked that in the case of stellar photometry, if whole star profiles are being compared to each other, the error due to the aperture effect becomes of second order.

The aperture effect described above must be considered a disadvantage for electronography when compared with photography since in measuring the density on a photographic plate one is using the logarithm of exposure, which more closely resembles the response of the measuring machines.

It has already been pointed out that the sensitivity of electronographic photocathodes varies both continuously and discretely. It was also remarked that it is in principle possible to correct for continuous variations in sensitivity. To do this, use is made of the fact that at every point on the photocathode photographic density is proportional to exposure, although the constant of proportionality varies over the surface of the photocathode. In order to effect a calibration it is necessary to obtain an exposure of the uniformly illuminated photocathode. The resultant densities may then be taken as the constant of proportionality at each point. To calibrate an exposure it is then merely necessary to divide the density at each measured point by the corresponding density from the calibration exposure. This process is called flat-fielding.

The preparation of an accurate photocathode map is a matter of critical importance in electronographic photometry, and it is an area which has received insufficient attention to date. Changes in the observed sensitivity pattern are produced in two different ways: inherent variations in the sensitivity of the photocathode caused by the manufacturing process, and variations in the thickness or transparency of the mica window. In principle these two patterns may move relative to each other on the film due to drift in the focusing system, but tests have shown that this drift does not normally exceed about 20 μm, and this is very small compared with the scale on which the sensitivity varies. The two effects will therefore be considered together.

The sensitivity pattern is obtained in the first instance by exposing the photocathode to a uniform distribution of light. The light source has traditionally been the dawn or dusk sky; in fact, the dawn sky is better as an area free from bright stars can then be chosen while it is still dark. The optimum photographic density for the film is in the region 0.5–1.5 as microphotometers become progressively less reliable at specular densities greater than this. It is however worth taking exposures of different lengths so that a check on the linearity of the system can be maintained. The exposure should be at least 20 seconds to ensure that shutter opening and closing times are negligible. The sky however must be bright enough to ensure that any faint stars in the field will not affect the sensitivity pattern. A second approach is to point the telescope at the inside of the telescope dome and use reflected light as a source of illumination. A translucent diffusing mask may be fixed over the telescope aperture, and this combined with the fact that the telescope is very far out of focus, produces uniform illumination.

Another type of photocathode non-uniformity appears in the form of dead spots or spots of zero sensitivity, normally formed when the photocathode is made, which are anything from a few microns to a few hundred microns across. There is of course no way of calibrating them out directly, although it is often possible to avoid them by arranging for objects of interest to fall on unaffacted parts of the photocathode. However, in the case of crowded fields or extended images this is not normally feasible. The effect of dead spots on an image will not necessarily be serious. For example, the general structure of the isophotes of a large image are unlikely to be affected by a small blemish. On the other hand, photometry is not possible under these circumstances. The best way of dealing with this situation is to add several exposures of the same field, each slightly offset so that individual objects fall on different parts of the photocathode. This technique can also be used to improve the overall signal-to-noise, although accurate registration and a small sampling interval are necessary for it to be effective.

4.6 Advantages and disadvantages of electronography

The two main advantages of electronography over photography are the high quantum efficiency and the linearity of response. The photocathode increases the light gathering power of a system by an order of magnitude, making observations possible on a small telescope which have in the past been reserved for the prime focus of large reflectors. Linearity makes possible photometric observations with only a zero point calibration, since the measured density is directly proportional to the incident flux and no additional calibration curves are necessary. This linearity holds at all light levels and hence implies no reciprocity failure or decreased response at low light levels, a factor which seriously limits the efficiency of photographic emulsions in many astronomical contexts. The very fine grain of nuclear emulsions used in electronography brings a number of additional advantages, including high signal-to-noise and large storage capacity enabling long exposures to be made. Adjacency effects at points of high contrast are very small and usually not significant, giving accurate representation of image structure over a large density range and high density gradients. The sensitivity variations of the emulsion are generally small and the principal variation in response comes from the photocathode. This latter variation can be removed by a flat-field calibration.

One of the drawbacks of the Spectracon was the small size of the window, since this caused difficulties for some of the most attractive electronographic projects such as stellar photometry and isophotometry of galaxies. The McMullan camera with a 40 mm diameter window has essentially solved this problem, but has aggravated the associated difficulties of constructing large photocathodes with satisfactory uniformity without pinholes or other blemishes. Improvements have had to be made in the technique of processing photocathodes, in particular in the prevention of dust particles from falling on the substrate during processing. Photocathode blemishes are a serious nuisance in many types of work, and in particular make surface photometry unnecessarily complicated.

The quality of nuclear emulsion also presents a problem in practice, as it is frequently not manufactured to specification in various ways. Specifically, new film often arrives with dust embedded in it which cannot be removed without marking the emulsion. The area of film spoilt by dust and other blemishes can sometimes be as high as 50 per cent.

Phosphor tubes used with photographic detectors suffer from many of the difficulties of ordinary photography. However, due to the light amplification, reciprocity failure is less important than with a photograph, and the signal may be more easily brought on to the linear part of the response curve. On the other hand phosphor tubes do not have the large-field advantage of a photograph, their signal-to-noise is usually poor and there is no advantage in storage capacity. Image definition is typically poor and in addition photocathode blemishes cause problems, as with electronography. In short, phosphor tubes used photographically have essentially no advantage over electronographic cameras, and a number of disadvantages.

Electronography compares unfavourably with solid state detectors in certain respects. The main, and for many observations overriding, advantage of electronography is the relatively large field obtainable. How long it will be before solid state detectors can be made with a large format remains to be seen, but for the present they are typically limited to perhaps 512 × 512 pixels. When a large field is not required, solid state detectors have a number of advantages over electronography. The process of development and measurement is eliminated, and in many cases real-time display of the data is possible. Problems of emulsion and photocathode faults are offset by equivalent difficulties with all types of solid state detector, which either have photocathodes as well, or in the case of CCDs and similar devices are degraded by missing pixels and non-uniformity of response.

4.7 Electronographic observations

It would not be useful or feasible to attempt to give here an exhaustive list of the types of observation suitable for electronography. Successful applications must clearly take into account the advantages and disadvantages of the method. This consideration has on occasions been overlooked in the past, and some early work was attempted in situations where electronography was not the best approach, leading to some reservations on the part of astronomers as to its overall usefulness.

By far the most successful application of electronography to date has been in stellar photometry. The wide field of modern electronographic cameras is sufficient to obtain large samples of stars, and for most purposes images affected by dead spots can simply be omitted, or re-observed on a different part of the photocathode. The procedure for making the observations is fairly straightforward. A number of exposures are obtained of the field in the wavebands required. Programme stars and a few photoelectric standard stars are then measured as small raster arrays using a microphotometer. The density in each image is then integrated either directly or by fitting a profile, typically a Gaussian. The resulting value must then be corrected for the variation in photocathode response from previously measured and processed photocathode sensitivity maps. The relative flux and hence magnitude difference is then obtained directly on the basis of the linear relation between density and flux, and converted to true magnitude by the addition of a zero point from the known magnitudes of the standard stars. In practice, a few additional complications may be involved. If high densities have been used, small corrections may be necessary for non-linearity in both the microphotometer and the emulsion. The colour response of the photocathode/filter combination must be corrected to that of the standard system. If the available photoelectric standard images are saturated on the electronographs, intermediate exposures may be necessary to bridge the gap, such that the standards are not saturated and yet there is an overlap of stars which can be measured on both short and long exposures. The typical safe dynamic range for one exposure is about four magnitudes. In addition there are a number of ways in which the measurements of the images may need to be processed further, for example by removing or replacing apparently spurious pixels, but discussion of this is outside the scope of the present work.

From the start, the idea of using electronography for galaxy photometry has appeared very attractive, since the linear response avoids the difficult procedure of calibrating isophotal maps with calibration curves. However, there are also drawbacks in using electronography for this purpose. Until the introduction of the McMullan camera, the field available on the Spectracon was not really large enough to allow useful work to be done to compare with other methods. A more difficult problem lies in the effect of the dead spots and blemishes, which cannot usually be avoided. To obtain a reliable map it is necessary to take several exposures of the object on different parts of the photocathode, align the measurements, normalise and add pixel-by-pixel. This is a difficult procedure and few satisfactory results have so far been published. Another difficulty which is particularly relevant for measuring low surface brightnesses is to obtain a sufficiently accurate flat background. It is frequently desirable to measure faint extensions perhaps only 2 per cent as bright as the night sky, and this puts very severe constraints on the accuracy of the flat-fielding procedure. Undoubtedly these problems will be overcome in the future, but to date galaxy isophotometry using electronography has not proved very successful.

The ability of an electronographic camera to record an image profile with high accuracy makes some interesting work possible in the field of image analysis. In particular it has been possible, using Fourier techniques and other methods, to take the images of double stars and other objects that have been blurred by atmospheric seeing and deconvolve them to a resolution several times better than the seeing disk. This requires excellent signal-to-noise, but is a field of research which holds considerable promise for the future.

References for Chapter 4

Heudier, J.L. & Sim, M.E. (eds.), 1981. *Astronomical Photography 1981* – Proceedings of meeting of International Astronomical Union Working Group on Photographic Problems held in Nice, April 1981.

James, T.H. (ed.), 1977. *The Theory of the Photographic Process*, 4th edition, Macmillan Company.

Marton, L. (ed.), 1966. *Advances in Electronics and Electron Physics*, Vol. 22A, Academic Press.

McMullan, D. & Morgan, B.L. (eds.), 1978. *Proceedings of Seventh Symposium on Photoelectronic Image Devices*, Imperial College, London.

5

Photoconductive detectors

The majority of electronic detectors used in modern astronomy have as their photon detecting element a semiconductor photocathode or target. In order to understand the factors defining the quantum efficiency, spectral range and sensitivity of these detectors, we need to consider the basic physical processes involved in the operation of a semiconductor and its interaction with radiation. The most fundamental process is that of photoconductivity, which forms the subject of this chapter.

5.1 Intrinsic semiconductors

In general, all solids can be classified as conductors, insulators or semiconductors, although, strictly speaking, semiconductors are a special class of insulators. The electrical behaviour of solids under the influence of electric fields, electromagnetic radiation and heat is conveniently represented by the energy band diagram, in which electron potential energy is plotted against distance through the solid. Figure 5.1 shows representative energy band diagrams for the three classes of solid. The lower energy band is that containing the outermost shell or valence electrons and hence called the valence band. The uppermost or conduction band, which may be either empty or partially filled, represents those electrons which are free to move through the solid. In conductors the valence and conduction bands overlap so that free electrons are available for current flow whereas insulators and semiconductors possess an energy gap, or forbidden band, separating the valence and conduction bands. Electrons may be excited from the valence to the conduction band through the addition of thermal or electromagnetic energy. The excitation of electrons to the conduction band is, of course, the basic operating principle of all optical semiconductor detectors, whereas the thermal effect is the major source of noise. The difference between insulators and pure, or intrinsic, semiconductors is due solely to the respective sizes of energy gap. At absolute zero the conduction bands of both insulators and semiconductors are empty. However, above this temperature there is a finite probability that the thermal energy in the lattice will cause one of the valence electrons to be freed and move into the conduction band, leaving behind a positively charged ion or 'hole' in the valence band. The movement of holes throughout the solid does not imply the physical movement of positive ions but refers to the continuous production and neutralisation of positive ions as electrons move through the solid from ion to ion. The movement of electrons in one direction is equivalent to the movement of holes in the other direction.

Also shown in Figure 5.1 is the Fermi level, E_F, which is the energy at which the probability of the corresponding electron energy state being occupied by an electron is one-half. The laws of statistical mechanics

Figure 5.1. Energy band diagrams in which electron energy is plotted against distance through the material. The diagrams demonstrate the essential difference between conductors, semiconductors and insulators.

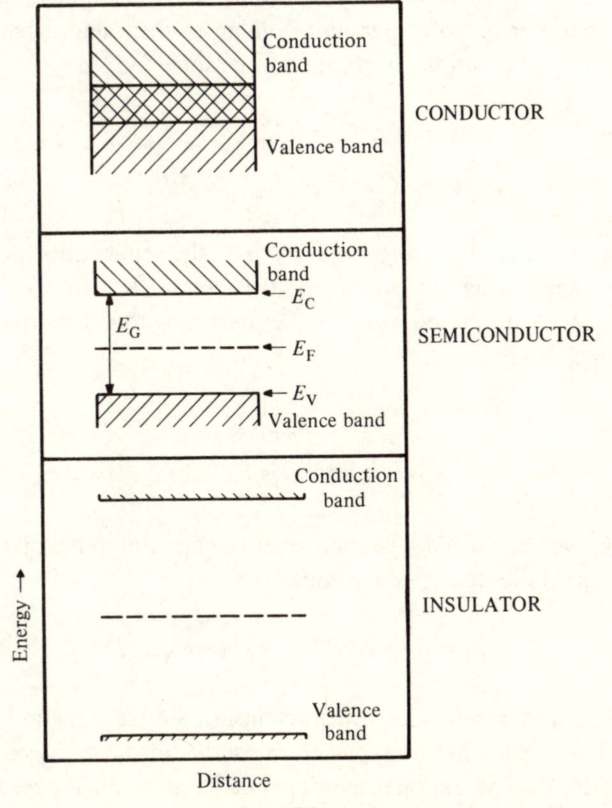

govern the behaviour of electrons in solids, and show that the probability, f, of any other energy state, E, being occupied is:

$$f = \{\exp[\,(E - E_F)/kT] + 1\}^{-1},$$

where k is Boltzman's constant and T is the absolute temperature of the solid. For an intrinsic semiconductor, E_F is situated exactly halfway between the top of the valence band, E_V, and the bottom of the conduction band, E_C, at absolute zero, and only slightly depressed from centre at 300 K. The importance of the Fermi level is that solids in field-free thermal equilibrium have a constant Fermi level throughout the solid. Furthermore, two solids in contact with each other adjust their electron–hole populations so that the Fermi level is the same throughout both solids. The movement of charges across the solid junction required to redistribute the electron–hole populations is the chief factor in determining the junction's electrical characteristics.

The number of electrons per unit volume in the conduction band, n, and the number of holes per unit volume in the valence band, p, are related to E_C, E_V and the absolute temperature by:

$$n = AT^{3/2} \exp[(E_F - E_C)/kT],$$

$$p = BT^{3/2} \exp[(E_V - E_F)/kT],$$

where A and B are constants depending on the semiconductor material, and are approximately equal to 10^{-24} mm^{-3} K$^{-3/2}$ for silicon and germanium. Multiplying the two expressions together removes the dependence on E_F:

$$np = ABT^3 \exp[(E_V - E_C)/kT],$$

or

$$np = ABT^3 \exp(-E_G/kT),$$

where E_G is the forbidden band energy gap. For pure materials the electron and hole densities are equal:

$$n_i = p_i = (ABT^3)^{1/2} \exp(-E_G/2kT)$$

where the suffix, i, refers to pure or intrinsic material. For good insulators E_G is at least 5 eV whereas typical semiconductors have E_G values of the order of 1 eV. The exponential dependence of n_i on E_G results in the conduction band in insulators being essentially empty at room temperature whereas there is a measurable free electron population in intrinsic semiconductors at room temperature.

The electron (and hole) population is not constant but fluctuates because some electrons recombine with holes and return to the valence band while at the same time new electrons are being continuously excited to the conduction band. Furthermore even so-called pure semiconductors contain both lattice defects and states associated with the crystal surfaces. These can have a large effect on the behaviour of the semiconductor by providing centres at which electrons and holes can be trapped and recombine.

The energy gap, E_G, is itself weakly dependent on temperature; the exact dependence is a function of the semiconductor material. For all astronomically important semiconductors, E_G increases with decreasing temperature. As an example, the energy gap in silicon increases from 1.12 eV at 300 K to 1.17 eV at 0 K.

The semiconductors most commonly found in the various solid state detectors described in this and later chapters are two of the elements in group IV of the periodic table, silicon and germanium, in which the crystal lattice is formed from atoms sharing four valence electrons with four surrounding atoms, as shown in Figure 5.2. The energy gap at 300 K for germanium is only 0.68 eV, with the result that germanium has about 1000 times the conductivity of silicon at room temperature. Since thermally-generated conduction-band electrons generally represent 'noise', the advantage of silicon over germanium for uncooled detectors is obvious. The exponential dependence of n on temperature means, for

Figure 5.2. Two-dimensional representation of the lattice structure in a silicon crystal. Each pair of lines represents a covalent bond linking adjacent silicon atoms.

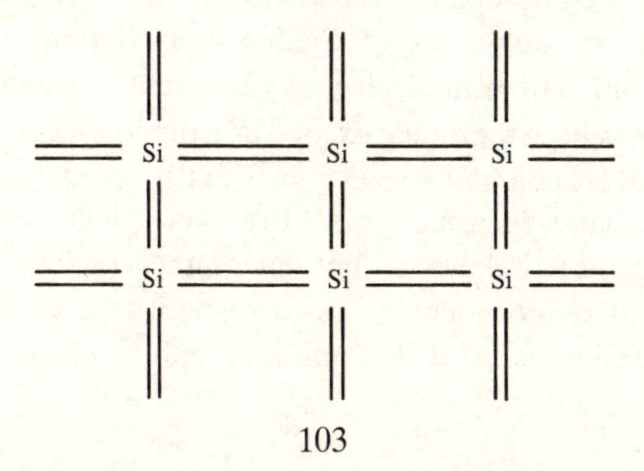

silicon, a halving of the thermal noise for every 10 K decrease in temperature. For germanium the same temperature decrease reduces the thermal noise by about 30 per cent.

Another group of semiconductors, important as photocathodes, comprises the group III–V compounds, principally the arsenides, antimonides and phosphides of gallium and indium. In this group there is a wide range of energy gaps ranging from 0.18 eV for InSb to 2.25 eV for GaP. Another compound semiconductor, of importance in television type detectors, is one of the oxides of lead, PbO, which has an energy gap of 1.9 eV. Semiconductors can also be formed from group II–VI compounds but are not yet of importance to astronomy.

The significance of the various semiconductors is directly related to the size of their energy gaps and will be considered in some detail below. However, the effect of doping on semiconductor properties is discussed first.

5.2 Doped semiconductors

The behaviour of a semiconductor is changed by the process of doping, that is by introducing controlled amounts of impurity atoms into the crystal lattice. The properties of these doped, or extrinsic, semiconductors are significantly altered even though the level of impurity is typically of the order of only one part in 10^8. The importance of doping can be judged from the fact that virtually all electronic detectors used in astronomy rely on some form of doped semiconductor for their operation.

The simplest intrinsic semiconductors are the group IV elements silicon and germanium. An atom of the group V element arsenic is very nearly the same size as an atom of germanium and can be easily introduced into the crystal lattice to replace it. However, although arsenic has five valence electrons, only four of them will be bound to the nearby germanium atom. The fifth electron is only loosely bound to the parent atom and can easily be thermally excited into the conduction band leaving behind a positively charged arsenic ion. The fifth electron is represented on the energy band diagram by an extra energy level in the forbidden band about 0.1 eV below the bottom of the conduction band. The resulting doped semiconductor is said to be n-type with the group V element referred to as the donor impurity and the electrons designated the majority carriers.

In a similar fashion a group IV atom can be replaced by a group III atom which has only three valence electrons. In this case one of the four valence electrons of a nearby group IV atom can be thermally freed and then accommodated into the group III atom's outer shell. This process creates a negatively charged ion and leaves behind a positively charged group IV ion or 'hole'. As a result an extra energy level is created about 0.1 eV above the top of the valence band. This type of doped semiconductor is called p-type and the holes are called the majority carriers.

Compound (group III–V) semiconductors are doped by replacing either the group III or group V elements by a group IV element. An n-type semiconductor results from substitution of the group III element. Replacing the group V element gives a p-type semiconductor.

As would be expected the presence of these extra holes or electrons changes the electron–hole statistics and moves the Fermi level. For all practical doping levels the number density of electrons, n, in the conduction band almost exactly equals the number density of donor atoms, N_d. Similarly, the number density of holes, p, in the valence band almost exactly equals the number density of acceptor atoms, N_a. The expressions for charge concentrations can then be written:

$$n = N_d = AT^{3/2} \exp[(E_F - E_C)/kT],$$

$$p = N_a = BT^{3/2} \exp[(E_V - E_F)/kT],$$

noting that A and B depend on the particular semiconductor material. If these expressions are rewritten:

$$E_F - E_C = kT \log (N_d/AT^{3/2}),$$

$$E_V - E_F = kT \log (N_a/BT^{3/2}),$$

the dependence of Fermi level on doping can be seen. For an n-type semiconductor increasing the doping, N_d, moves the Fermi level upwards in energy towards the conduction band. This is to be expected as the Fermi level can be thought of as the mean electron level and would be expected to rise as the conduction band is filled. Conversely increased p-type doping lowers the Fermi level. Figure 5.3 shows the various energy levels for both types of semiconductors.

Figure 5.3. Energy band diagrams for n-type and p-type doped semiconductors. In the n-type case the donor atoms are ionised and release free electrons into the conduction band leaving positive donor ions just below the conduction band. In the p-type case the acceptor atoms capture electrons leaving positively charged holes in the valence band.

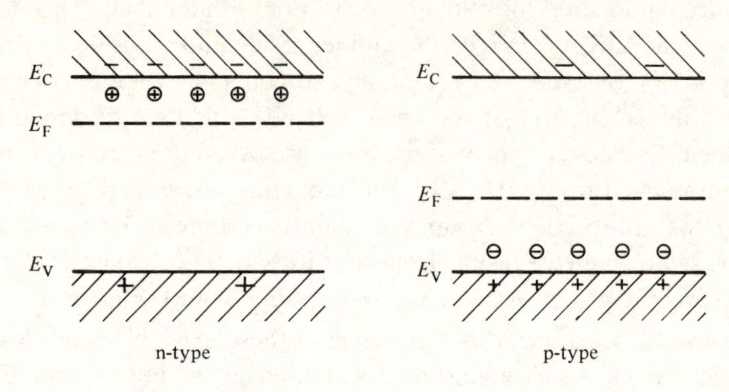

5.3 The photoconductive process

The intrinsic photoconductive process is the absorption of a sufficiently energetic photon by the semiconductor crystal lattice such that one of the covalent bonds is broken and an electron–hole pair is created. To be specific, for an electron–hole pair to be created the photon wavelength must be shorter than a threshold wavelength, λ_t, given by:

$$\lambda_t = hc/E_G,$$

where h is Planck's constant, c is the speed of light and E_G is the energy gap. Table 5.1 lists the long-wavelength threshold, λ_t, and energy gap, E_G, at room temperature (300 K) for several semiconductors with important astronomical applications.

Photons of higher energy (i.e. shorter wavelength) enable electrons deeper in the valence band to be excited into the conduction band. The cut off at photon energies less than E_G would be expected to be very sharp. However, there is a low energy tail due to two related effects. Charge pairs which are not swept to opposite ends of the semiconductor by an externally applied electric field will quite rapidly recombine with the crystal lattice. The surplus energy is absorbed in the form of mechanical vibrations which propagate throughout the lattice and are known as phonons. If the lattice absorbs another photon simultaneously with a phonon, the combined energy in the lattice might just exceed E_G. The

Table 5.1. Long-wavelength threshold, λ_t, and energy gap, E_G, at room temperature for electron–hole pair production in various semiconductors.

Semiconductor	E_G (eV)	λ_t (μm)
Si	1.12	1.10
Ge	0.68	1.82
GaAs	1.35	0.92
GaP	2.25	0.55
GaSb	0.78	1.59
InAs	0.33	3.8
InP	1.25	1.00
InSb	0.18	6.9
PbO	1.90	0.65

second effect is the slight fluctuation in E_G caused by random thermal vibration which enables lower energy photons to create charge pairs. As mentioned previously, E_G is a weak function of temperature and thus λ_t is also a weak function of temperature. Cooling of the semiconductor increases E_G and hence shortens the cutoff wavelength. This can be particularly bothersome with near infrared detectors since cooling is essential to reduce thermal noise.

The creation of an electron–hole pair temporarily increases the local conductivity. To detect this externally a uniform electric field is established in the semiconductor crystal such that electrons and holes are swept to opposite ends of the crystal. This is accomplished by connecting a voltage in series with a load resistor, R_L, between the ends of the crystal as shown in Figure 5.4. The increase of conductivity is detected as an increase in the voltage developed across R_L which for optimum sensitivity should be equal to the detector resistance. In a practical photoconductor photons are incident on only one surface of the crystal. When a charge pair is created, the more mobile electrons diffuse away from this surface faster than the holes and so create a carrier population gradient with its associated electric field perpendicular to the longitudinal externally

Figure 5.4. (a) Circuit diagram showing an intrinsic photoconductor across which is maintained a voltage, V, in series with a load resistor, R_L. (b) Corresponding energy band diagram showing all energy levels tilted by V.

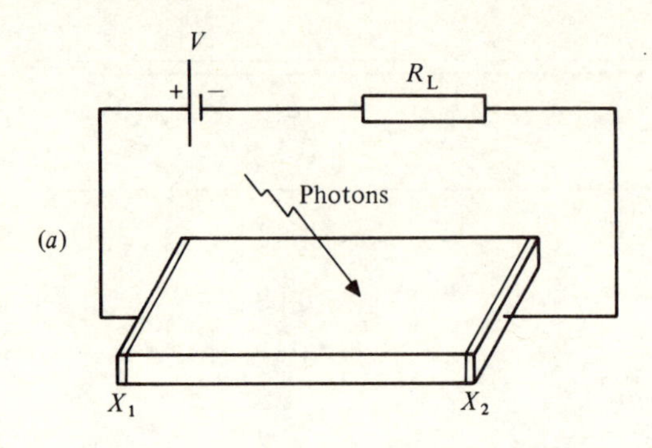

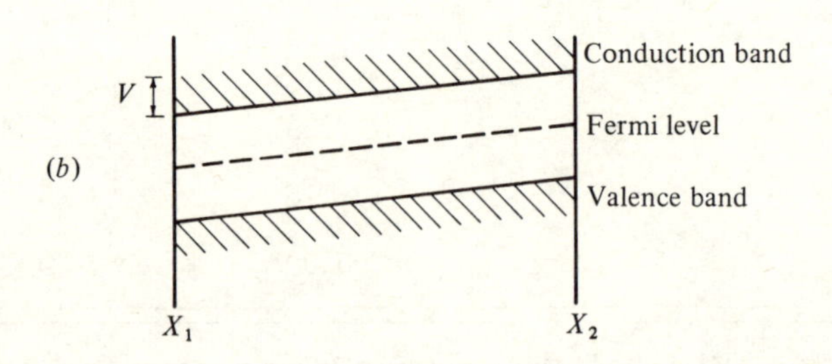

applied field. If the two fields become comparable, the diffusion of electrons will be inhibited and non-linearity will result. Also shown in Figure 5.4 is the corresponding energy band diagram. The applied voltage, V, tilts the energy bands uniformly throughout the crystal in a direction such that the energy levels at the more positive end of the crystal are depressed by V.

The term intrinsic photoconductor refers to those photoconductors for which the basic detection is that just described. The term also applies to doped semiconductors because the detection process still involves a property intrinsic to the bulk semiconductor material and not due to any impurities present, that is the excitation of an electron by a photon from the valence band to the conduction band. If doping is used, the type and level of doping affect both sensitivity and speed of response, with p-type usually preferred.

In an extrinsic photoconductor, impurity atoms are introduced into the crystal lattice such that additional energy levels are created within the forbidden energy band. If an incident photon has sufficient energy to ionise the impurity atom a free electron is added to the conduction band of the bulk semiconductor. The term extrinsic refers to the fact that the effective energy gap for long wavelength radiation is not a property of the bulk semiconductor itself but is controlled by the ionisation energy of the impurity added to the bulk semiconductor. In this way very long wavelength photons can be detected provided the crystal is cooled sufficiently so as to eliminate thermally induced transitions. For this class of photoconductor the sensitivity at low impurity concentrations is directly proportional to the impurity concentration. Table 5.2 lists the ionisation energy, E_i, and corresponding long wavelength threshold, λ_t, for some typical semiconductor–impurity combinations.

Table 5.2. Ionisation energy, E_i, and corresponding long-wavelength threshold, λ_t, for some typical semiconductor–impurity combinations.

Semiconductor–impurity	E_i (eV)	λ_t (μm)
Ge–In	0.011	113
Ge–Ga	0.011	113
Ge–As	0.013	95
Ge–P	0.012	103
Si–In	0.16	8
Si–Ga	0.068	18
Si–As	0.053	23
Si–P	0.045	28

5.4 The junction diode

The most commonly encountered junction diode is the simple p-n diode which is manufactured by differentially doping a homogeneous semiconductor crystal. There is a very narrow region in which the doping changes from p-type to n-type known as the p-n junction. This region determines the essential behaviour of the p-n diode. The processes occurring at the junction can best be understood by considering the p-type and n-type semiconductor as two physically separate semiconductors brought into contact. Upon contact the holes on the p-side of the junction and the electrons on the n-side of the junction will tend to diffuse across the junction. In doing so the positively charged donor ions and the negatively charged acceptor ions will be left behind, giving rise to an electric field that opposes any further diffusion across the junction. The electrons and holes in the vicinity of the junction will recombine to leave a charge-free depletion layer. The potential across this depletion layer is just that necessary to shift the energy bands to keep the Fermi level constant throughout the whole semiconductor. As shown in Figure 5.5, the conduction and valence bands are bent in the vicinity of the junction. This electron field is confined to the depletion layer owing to the relatively high resistivity of this region compared to elsewhere in the semiconductor. The potential drop across the depletion layer is 0.6 V for silicon and 0.3 V for germanium. As a result, since electrons must have an energy greater than this potential to cross the junction, germanium has a higher leakage current than silicon.

If a potential difference, V, is applied across the diode such that the p-side is made positive with respect to the n-side, the diode, or junction, is said to be forward biased, as shown in Figure 5.6. Because of the high resistivity of the depletion layer, most of the potential V appears across the depletion layer with the energy levels remaining virtually 'horizontal' elsewhere.

The applied electric field opposes the intrinsic field so that leakage across the junction increases. In fact if V equals the intrinsic voltage

Figure 5.5. Energy band diagram for an unbiased p-n junction. Charge pairs are removed from the depletion region until the Fermi level is the same on both sides of the junction.

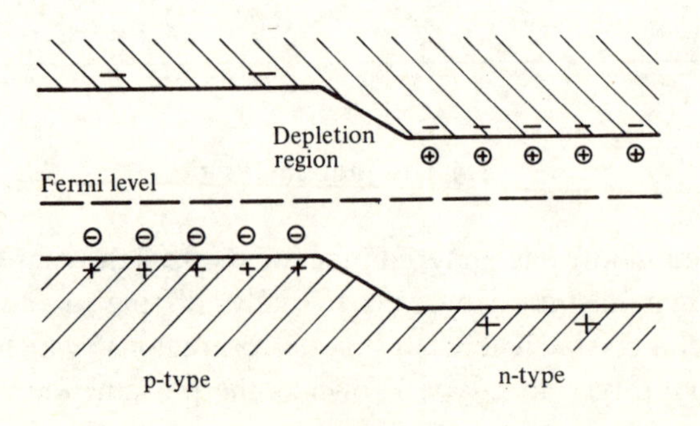

Figure 5.6. Energy band diagram for a forward biased p-n junction. The Fermi level in the n-type semiconductor is raised relative to the level in the p-type semiconductor by an amount equal in electron volts to the externally applied bias voltage, V.

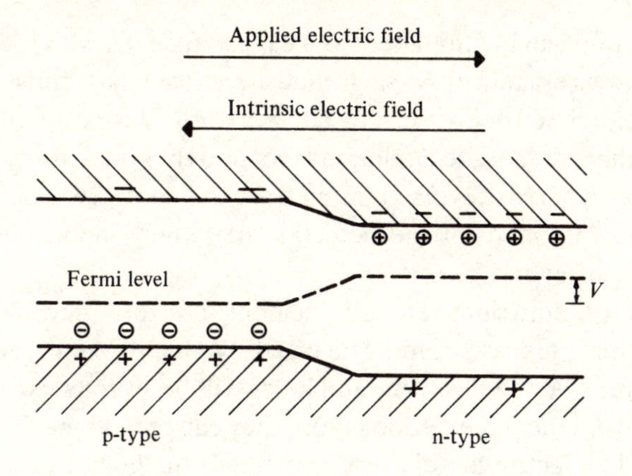

Figure 5.7. Energy band diagram for a reverse biased p-n junction. In this case the Fermi level in the n-type semiconductor is lowered by V electron volts.

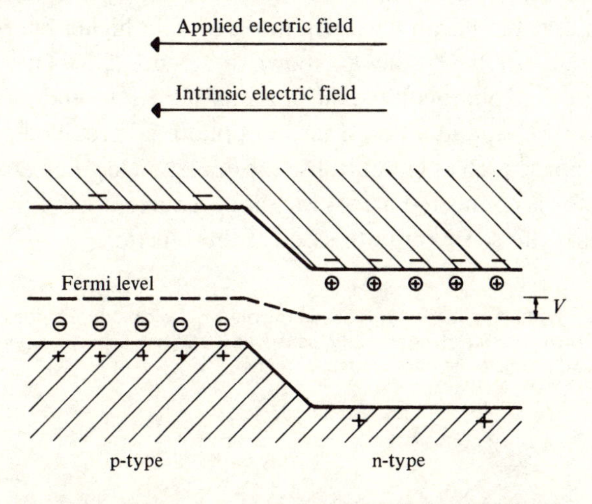

(0.6 V for silicon) electrons flow unimpeded across the junction with the current limited only by the external circuit.

Applying the external potential, V, across the diode in the opposite sense reverse biases the diode creating a higher barrier to leakage current, as shown in Figure 5.7. From this figure, it can be seen that at a sufficiently high reverse bias, valence band electrons on the p-side of the junction will have the same energy as unfilled energy states in the conduction band on the n-side. Under this condition tunnelling can occur directly across the junction. This phenomenon is called Zener breakdown, and it limits the maximum reverse voltage in diodes to a few tens of volts.

The junction can be thought of as a capacitance, C, with the rest of the diode a series resistance, R, such that there is a characteristic RC time constant associated with the device. Forward biasing of the junction decreases the width of the depletion layer and thus increases the junction capacitance, whereas reverse biasing does the contrary. For this reason most diodes intended for the detection of rapidly modulated light are operated with reverse bias.

Photons of sufficient energy incident on the diode will create electron–hole pairs as before. The electric field present in the depletion layer ensures that the electrons and holes will be swept away in opposite directions from the p-n junction before they can recombine. Elsewhere in the diode the electric field is very low due to the higher conductivity so that any charge pairs will have a high probability of recombination.

In the absence of any external load or bias across the diode there will be a buildup of electrons on the n-side of the depletion layer and holes on the p-side such that the Fermi level will move up to a higher energy on the n-side relative to the p-side as shown in Figure 5.8. This mode of operation is called photovoltaic; the potential across the ends of the diode is proportional to the number of incident photons, provided the charge collected is not so high as to neutralise the depletion layer electric field. If a load resistor is connected across the diode a current will flow until the Fermi level is the same on both sides of the junction.

Figure 5.8. A p-n junction diode operating in the photovoltaic mode. Photons which enter the depletion region create electron–hole pairs which are swept apart to opposite sides of the junction and produce the photovoltaic potential, V.

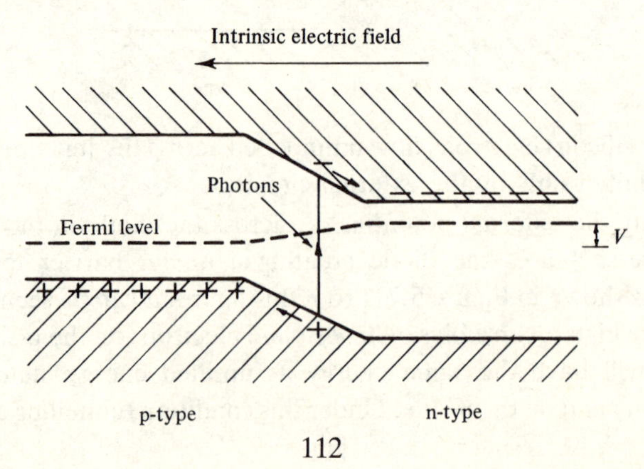

The normal mode of diode operation is with a reverse bias in series with a load resistor connected across the diode. In this photoconductive mode of operation charge pairs are removed from the region of the junction as soon as they are formed which leads to a highly linear response. Typical diodes are linear over eight decades of incident photon flux. The presence of the reverse current gives rise to a leakage or dark current which has associated with it several sources of noise. For this reason, some astronomical uses call for unbiased operation but use a load resistor, thereby operating in the highly linear photoconductive mode rather than the photovoltaic mode which would be the case if the diode were open circuited. As the most important diodes are all intrinsic photoconductors, the optical wavelength response is the same as that for homogéneous photoconductors with the long-wavelength thresholds given in Table 5.1.

The inclusion of an intrinsic or undoped layer of semiconductor between the p and n doped layers results in a p-i-n diode such as shown in Figure 5.9. The high resistivity of the intrinsic region means that the junction electric field extends throughout the region and is equivalent to a greatly increased depletion layer. This increases the probability of capture of incoming photons particularly at the longer wavelengths of the near infrared. Furthermore the decreased diode capacitance as compared to the p-n diode results in a faster response time and a more sensitive detector in practice.

The application of metallic contacts to the semiconductor can change the band structure at the edge of the semiconductor crystal. If the metal–semiconductor junction were perfect the energy bands would be

Figure 5.9. A reverse biased p-i-n photodiode showing the greatly increased depletion region caused by the intrinsic region.

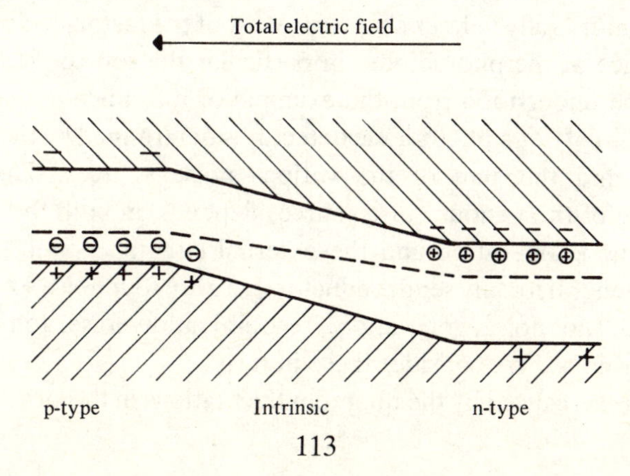

horizontal right up to the edge, as is shown in Figure 5.10 for the ohmic junction. However, in most practical cases there are defects associated with the edge of the semiconductor such that a certain fraction of conduction band electrons are trapped within the forbidden band. In an isolated semiconductor this would result in the depression of the Fermi level at the edge. Over a field-free junction, however, the Fermi level must remain constant and so the conduction and valence bands are forced to rise at the edge of the semiconductor. This type of metal–semiconductor junction is known as rectifying and is also shown in Figure 5.10.

Figure 5.10. Ohmic and rectifying junctions between metals and n-type semiconductors. Note the bending of the conduction and valence bands in the rectifying junction which is caused by imperfect surface states in the semiconductor.

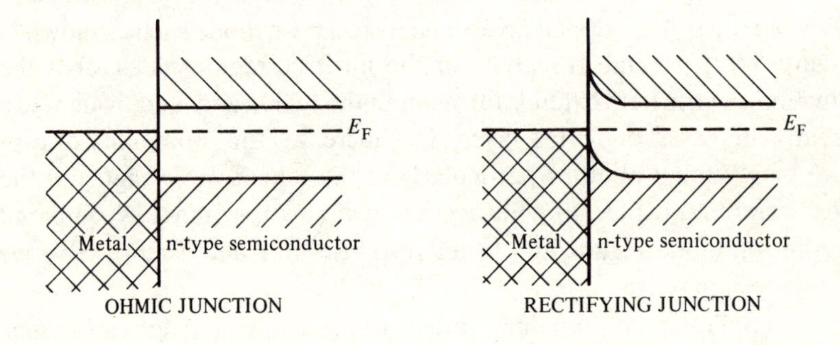

5.5 Practical photodiode operation

In astronomy, photodiodes are used either singly or, more commonly, in arrays such as the silicon vidicon target and the Reticon, both of which are described more fully in later chapters. Most of the factors which limit the performance of the photodiode, in particular the sources of noise, can however be understood from the example of the single diode.

The sensitivity of a photodiode detector is determined by the quantum efficiency and the sum of the various noise sources. The relative importance of the various noise sources depends on both the particular mode of diode operation and the external circuitry connected to the diode. In general for any semiconductor there are four noise sources to be considered, shot noise, generation–recombination noise, Johnson noise and flicker noise. We consider these in turn.

Shot noise is caused by the quantum fluctuations in the current flowing through the semiconductor device. The general form of the shot noise current is:

$$i_N = (2qIB + 4qI_oB)^{1/2},$$

where q is the charge on the electron, I is the diode current, I_o is the reverse bias or leakage current and B is the measurement frequency bandwidth. Most photodiodes are operated reverse biased so that the total diode current, in the absence of photon induced current, is simply the leakage current, I_o:

$$I = -I_o,$$

and

$$i_N = (2qI_oB)^{1/2}.$$

Generation–recombination (G–R) *noise* is caused by thermally generated fluctuations in the rates of carrier generation and recombination which in turn cause a fluctuation in the semiconductor resistance and hence a fluctuation in the voltage across the semiconductor when a bias current flows. For n-type semiconductors, the equivalent G–R noise current is:

$$i_N = \frac{2I_o}{N}\left(\frac{PB\tau}{1 + (2\pi f\tau)^2}\right)^{1/2},$$

where P is the total number of free holes in the semiconductor, N is the total number of free electrons in the semiconductor, τ is the electron lifetime and f is the frequency at which the noise is being measured. For typical silicon p-n diodes this noise is much lower than the shot noise and so need not be considered further.

Associated with any resistance is *Johnson noise* which is due to Brownian motion of the charge carriers and is solely a function of temperature and measurement bandwidth. The equivalent Johnson noise current in a resistance R is given by:

$$i_N = \left(\frac{4kTB}{R}\right)^{1/2},$$

where k is Boltzman's constant, T is the absolute temperature and B is the measurement bandwidth. The equivalent bulk parallel resistance of modern p-n photodiodes is so high (about 10^{11} ohm) that this noise source need only be considered if the diode is operated at zero bias current, when the shot noise is zero.

Finally, *flicker noise* is associated with all electronic devices and has the general form:

$$i_N \propto I_o B^{1/2} f^{-1}$$

so that it is often referred to as '$1/f$' noise. For silicon diodes, van der Ziel maintains that fluctuations in the recombination current in the space-charge region are the main cause of the noise. However, it is also known that the form of construction and manufacture of the diode can affect the magnitude of the noise. A photodiode can usually be selected with a flicker noise component less than the shot noise. For very low frequency or d.c. operation, however, the flicker noise can become significant.

Both flicker and shot noise depend on bias current and lowest noise operation would be expected for zero bias. However, the equivalent parallel diode resistance, R_d, is also a function of reverse bias having a minimum resistance and hence maximum Johnson noise at zero bias. Lowest overall noise operation is actually obtained at a reverse bias of a few tenths of a volt.

Figure 5.11 shows a typical diode detector circuit in which an operational amplifier is employed as a current-to-voltage converter. The operational amplifier has both current and voltage noise sources chiefly given by the field effect transistor (FET) input stage usually employed in low noise amplifiers. The voltage noise source has $1/f$ characteristics up to

Figure 5.11. Photodiode detector used with an operational amplifier connected as a current-to-voltage converter. The photodiode is reverse biased. R_f is the external feedback resistor and C_f is the external feedback capacitor.

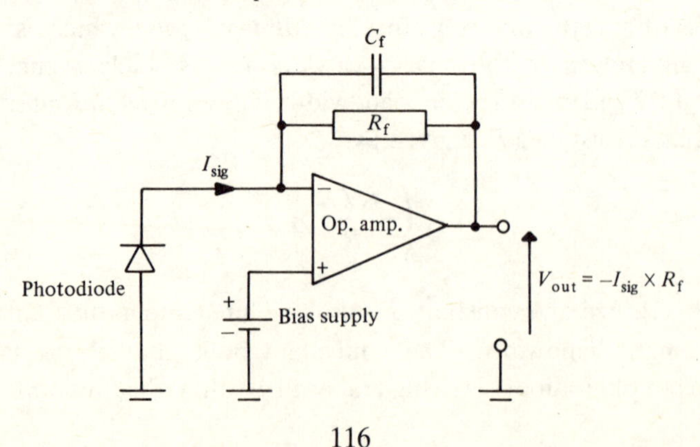

a frequency of a few hundred hertz and then stabilises at an approximately constant value up to several tens of kilohertz. This source of noise is not reduced by cooling and is the main limitation to performance in practical detector systems. The current noise source is simply the shot noise associated with the bias current flowing in the input FETs and can be almost eliminated by cooling. The feedback resistor, R_f, introduces Johnson noise and for this reason is made as large as possible and also cooled. The actual value of feedback resistor used depends on both the sensitivity and frequency response desired. The output voltage is related to the input current according to

$$\text{output voltage} = -\text{ input current} \times R_f$$

However, this relationship is only true for a steady input current. If the current is varying or modulated the output voltage will be lower and in fact will be reduced by 0.707 at a modulation frequency of

$$f = (2\pi R_f C_f)^{-1}.$$

As the feedback capacitance, C_f, is typically 10^{-12} farad, and response

Figure 5.12. Spectral response curve for a typical silicon p-i-n diode. The quantum efficiency approaches 80 per cent at 0.85 μm.

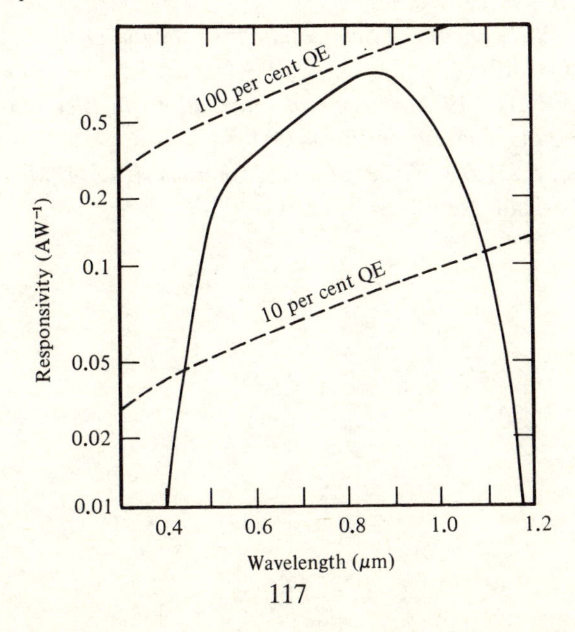

up to several tens of hertz is often required, R_f is usually of the order of 10^9 ohm. Operation at several tens of hertz is desirable in that flicker noise is reduced by at least an order of magnitude over d.c. operation.

In summary, the lowest noise operation occurs for a cooled small area (to give minimum leakage current) diode operated at a few tenths of a volt reverse bias connected to a cooled operational amplifier selected for lowest voltage noise and a cooled feedback resistor made as large as possible.

The spectral response of a typical silicon p-i-n diode is shown in Figure 5.12. The quantum efficiency (QE) can be seen to be approximately 70 per cent over the 0.5 μm to 1.1 μm spectral range thus making the diode useful for the *V*, *R* and *I* spectral bands. The long-wavelength cutoff is that described previously whereas the short-wavelength cutoff is due to absorption of photons before reaching the depletion region. The combination of quantum efficiency and noise sources yields a minimum detectable photon flux of about 2000 photons per readout.

References for Chapter 5

Hyde, F.J., 1965. *Semiconductors*, Macdonald & Co., London.

Kingston, R.H., 1978. *Detection of Optical and Infrared Radiation*, Springer-Verlag, Berlin.

Kruse, P.W., 1977. In *Optical and Infrared Detectors*, ed. R.J. Keyes, Springer-Verlag, Berlin.

Leck, J.H., 1967. *Theory of Semiconductor Junction Devices*, Pergamon Press, Oxford.

Robinson, F.N.H., 1974. *Noise and Fluctuations in Electronic Devices and Circuits*, Clarendon Press, Oxford.

van der Ziel, A., 1970. *Noise, Sources, Characterization, Measurement*, Prentice-Hall, New Jersey, USA.

6

Photoemissive detectors

The previous chapter has been concerned with the fundamental detection process in which sufficiently energetic photons create electron–hole pairs within the detector material. At very low light levels the photon induced electron current in the detector will be lower than the noise of any electronic amplifier connected to the detector. For faint object work the detector must employ some gain mechanism, the most usual being multiplication within the detector of those electrons produced by a photoemissive element or photocathode.

The photoemissive process in which photons release electrons (usually called photoelectrons) from a solid into a vacuum is extremely important in a wide variety of astronomical detectors such as the electronographic camera described previously in Chapter 4, the television-type detectors to be described in Chapter 7 and the photomultiplier tube. The photomultiplier tube has had a major impact on astronomical photometry and will be discussed in this chapter following a review of the more important photocathodes.

6.1 The photocathode

The photocathode is a particularly important part of any detector system since it establishes the maximum quantum efficiency and spectral bandwidth of the overall system. An understanding of the basic photoemissive process will help in appreciating the wide variety of photocathodes commercially available today. Photoemission can be thought of as comprising three consecutive steps, the absorption of a photon producing an electron–hole pair in the solid material, the motion of the electron towards the material–vacuum interface, and the overcoming of the potential barrier at the interface and the release of the electron into the vacuum.

Metals make poor photocathodes for the range of photon energies found in optical astronomy because most incoming photons are reflected at the surface rather than absorbed in the bulk material. Furthermore, any photoelectrons produced rapidly lose energy in collisions with the large number of free electrons present in any metal and the potential barrier to be overcome in escaping to the vacuum is several electron volts. On the other hand, semiconductors can be very efficient at absorbing photons, have very few free electrons and can be made with very small surface potential barriers. Figure 6.1 is an energy band diagram for a semiconductor modified to incorporate the solid–vacuum interface. The energy levels E_C, E_V, E_F and E_G are as defined in the previous chapter. The potential barrier to be overcome at the semiconductor surface is E_A, the electron affinity, and E_{vac} ($= E_C + E_A$) represents the energy of free electrons in the vacuum. The sum of the band gap energy and the electron affinity is the minimum incident photon energy required for electron emission. Since the electron affinity is generally greater than zero, incident photon energies must be higher in photoemissive detectors than in equivalent photoconductive detectors and thus photoemissive detectors have a generally poorer red response. Furthermore, photoemissive quantum efficiencies are typically only about 20 per cent as compared to the 80 per cent figure for photoconductive detectors because of the low probability of electrons reaching the semiconductor surface without suffering collisions. In photoconductive detectors there is a high probability of the electrons being detected in the bulk of the material.

It has been found possible to lower the electron affinity by strongly p-doping the semiconductor to bring the Fermi level down and then coating the semiconductor with a thin layer of an electropositive metal such as caesium. This creates a dipole electric field at the surface which bends the energy bands downwards until E_{vac} is only slightly higher than E_C, shown in Figure 6.2, and also extends the depth from which electrons can reach the surface. This process significantly improves both the photocathode red response and the quantum efficiency. In recent years this

Figure 6.1. Energy band diagram for a photoemitter showing the semiconductor–vacuum interface. E_{vac} is the energy electrons must possess in order to be released into the vacuum.

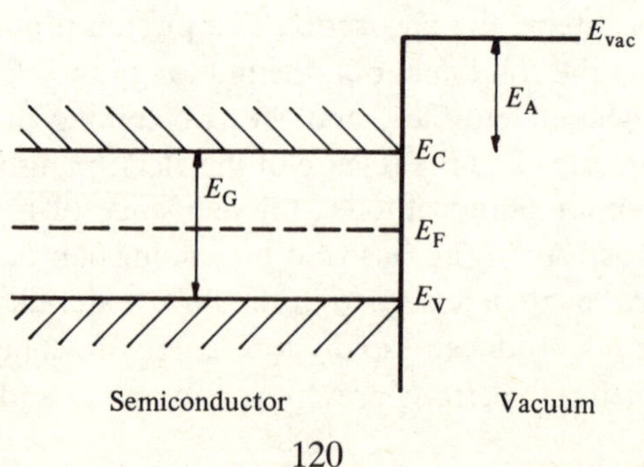

Figure 6.2. Energy band diagram for a coated semiconductor–vacuum interface. The electric field established in the metal coating causes the acceptor states near the coating to be filled. This results in energy band bending and a greatly reduced energy difference between E_C and E_{vac}.

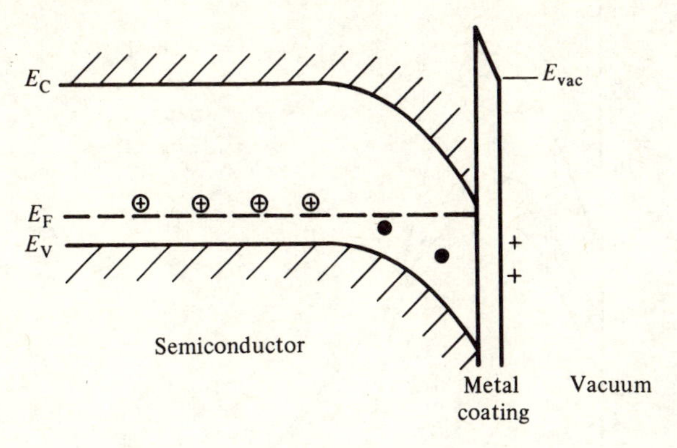

process has been taken a step further to produce negative electron affinity photocathodes by coating the semiconductor with a metal oxide, usually caesium oxide. In this case the long wavelength cutoff is simply that given by the band gap energy, E_G, as in the case of photoconductors.

There are about fifty different photocathodes commercially available. However, of these only a handful are generally used in astronomical applications. The more important photocathodes are listed in Table 6.1 and their spectral response curves shown in Figure 6.3. The oldest photocathode used in astronomy and still found in tubes such as the venerable 1P21 is caesium antimonide, Cs_3Sb, usually p-doped. The peak quantum efficiency is 22 per cent at 400 nm and due to the doping significant response extends to 700 nm. The size of this 'red tail' is somewhat dependent on the thermal history of the tube and care must be taken when doing broad-band filter photometry that photons longwards of 600

Table 6.1. Properties of certain photocathodes of importance in astronomy.

Photocathode type	Wavelength of peak sensitivity (nm)	Maximum quantum efficiency (per cent)	Thermionic emission at 20 °C (electrons s^{-1} cm^{-2})
S11	390	21	70
S20	380	22	300
S20-ER	400	20	400
Bialkali	380	27	15

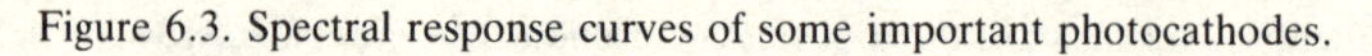

Figure 6.3. Spectral response curves of some important photocathodes.

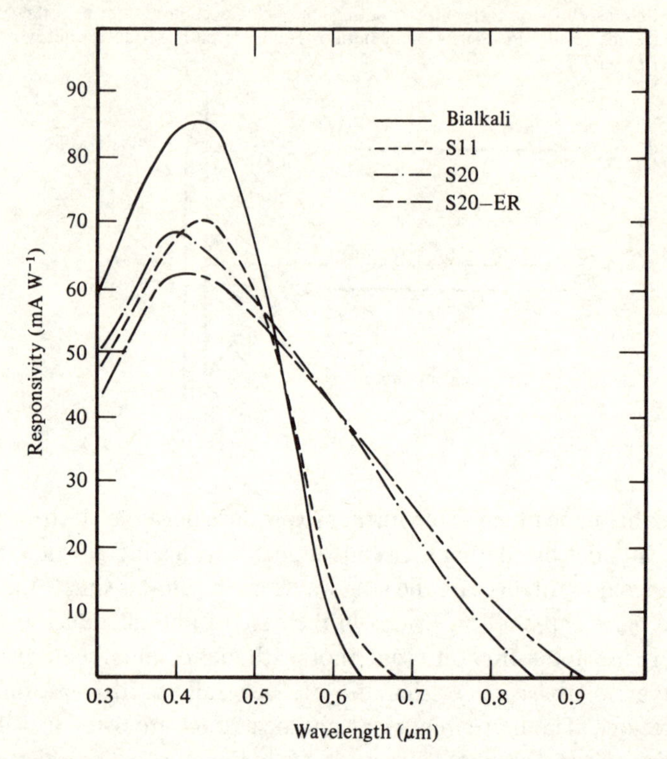

nm are not allowed to reach the tube, otherwise age dependent results may well be obtained. Photocathodes are usually given S numbers to designate their spectral response. Caesium antimonide may be used as an opaque photocathode, S4, or as a transparent one, S11. In an opaque photocathode the electrons are emitted from the same surface on which the photons impinge. In a transparent photocathode, strictly called semi-transparent, the electrons are emitted from the back surface. The most common photocathode material used today is the so-called trialkali compound $(Cs)Na_2KSb$, designated S20 and used in transparent photocathodes. The quantum efficiency at blue and near-UV wavelengths is the highest known and yet the useful response extends out to 800 nm. Photocathodes cannot be exposed to the air and are covered with protective windows. To obtain the blue response, a suitable UV-transmitting window material such as fused silica must be used in front of the photocathode. The caesiated photocathode enables electrons to reach the surface of the photocathode from deep within the material. Extended red response out to 900 nm at the cost of reduced blue response can thus be obtained by increasing the thickness of the photocathode. This is also shown in Figure 6.3. The bialkali photocathode, K_2CsSb, has been developed to give better response in the visible part of the spectrum than the S20 but has a greatly reduced red tail. The sharper red tail also means a much lower thermionic emission rate at room temperatures, although for most astronomical applications the whole photomultiplier tube is cooled. Since many different photocathode/entrance window combinations are now available it is possible to achieve a reasonable spectral match between the tube and the astronomical object being studied.

6.2 Photomultiplier tubes

The simplest photoemissive detector is the photodiode consisting of a photocathode and an anode enclosed in an evacuated envelope. Photons incident on the photocathode emit electrons (photoelectrons) which are collected by the anode and then amplified by external electronic circuitry. This device is useful for detecting high light levels but does not have the internal gain required to detect low light levels in the presence of amplifier readout noise. For this purpose the photomultiplier tube (PMT) has been developed. A typical photomultiplier tube configuration used for low light level photon counting is shown in Figure 6.4. Photons incident on the semitransparent photocathode cause photoelectrons to be released from the rear of the photocathode and to be directed by focusing electrodes onto the first dynode of an electron multiplier. The first dynode is maintained at a potential of a few hundred volts positive with

Figure 6.4. A typical photomultiplier tube showing the photocathode, dynodes and anode. This particular electrode configuration is known as linear focused.

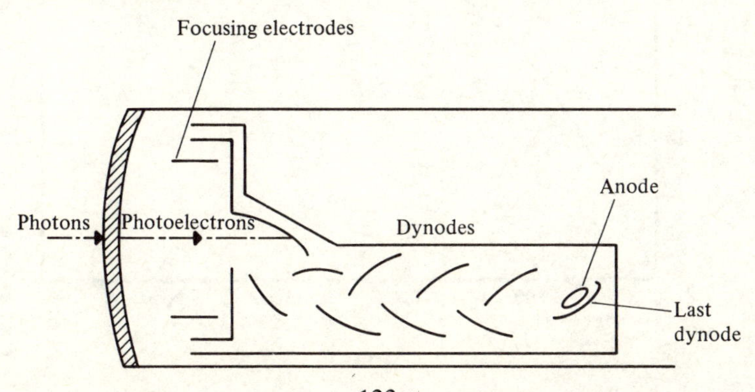

respect to the photocathode so that the photoelectrons strike the dynode with sufficient energy to release several secondary electrons per primary photoelectron. This process is repeated at successive dynodes so that the charge pulse leaving the last dynode will contain at least a million electrons for each primary photoelectron. A further electrode, the anode, situated close to the last dynode collects the charge pulse and passes it to the external circuitry. If the dynode secondary electron yield, defined as the average ratio of secondary electrons emitted per primary electron, is δ and the number of dynodes in the electron multiplier is N, the internal electron multiplier gain, G, is:

$$G = \delta^N.$$

The secondary emission of electrons is somewhat similar to the photoelectron emission from the photocathode. The primary electrons excite electrons within the dynode to higher energy states. Some electrons will be freed and travel to the surface of the dynode. Those electrons with sufficient energy remaining to overcome the electron affinity or work function of the dynode surface will be emitted into the vacuum carrying away any excess energy in the form of kinetic energy of motion. As would be expected, the secondary electron yield, δ, is a function of primary electron energy and Figure 6.5 shows the relationship for one of the more common dynode materials, MgO.

The number of excited electrons and the depth from which they are

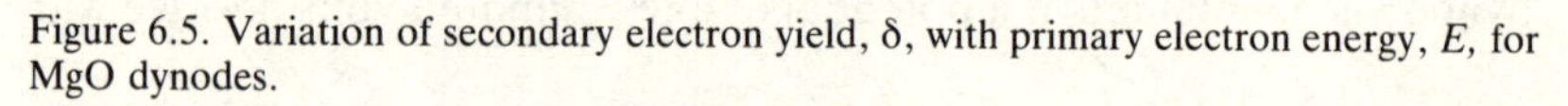

Figure 6.5. Variation of secondary electron yield, δ, with primary electron energy, E, for MgO dynodes.

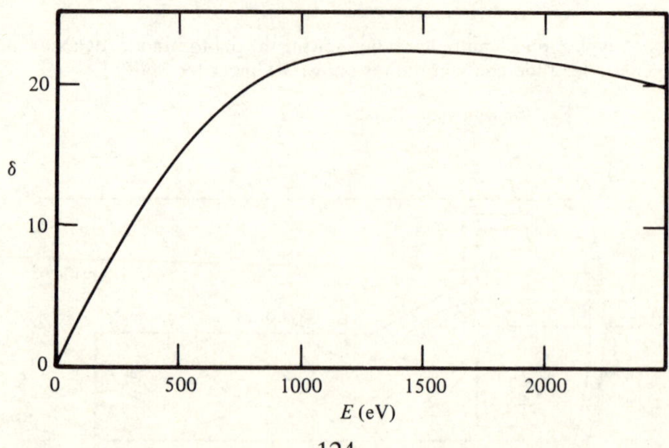

excited increases with primary electron energy. However if E is more than one thousand electron volts the excited electrons are produced at such depths in the dynode material that the probability of them reaching the surface becomes rather small. Recently dynodes have been made from negative electron affinity materials such as caesiated GaP. The negative electron affinity means that even those electrons which are produced deep within the dynode material and lose most of their energy by collisions on their way to the surface will still be released into the vacuum. For such materials, δ continues to rise with primary electron energies up to 10 keV at which point δ is about 200. Other common dynode materials are caesium antimonide, CsSb, and beryllium copper, BeCu.

The statistical nature of the electron excitement and transportation to the dynode surface causes the emitted secondary electrons to have a distribution of velocities which in turn causes a distribution in transit times between dynodes. This transit time distribution causes a smearing of individual charge pulses and sets an upper limit to the rate at which individual photons can be detected. Since the smearing out increases with the number of dynodes, it is preferable to use a few dynodes each with a high secondary electron emission coefficient. The anode is mounted close to the last dynode and often takes the form of a grid operated at a few hundred volts positive with respect to the last dynode. For fast time response the anode electrical impedance is designed to match the following amplifier input impedance, which is often 50 ohms.

Both semitransparent and opaque photocathodes are found in practical photomultiplier tubes. The popular 1P21 tube (shown in Figure 6.6) uses an opaque photocathode in a compact side viewing tube. Photons pass through the tube envelope and a metallic grill and then strike the photocathode releasing electrons which are accelerated to the nearby dynode. The compactness of the architecture gives a low electron transit time tube in a small physical volume and has proved to be a very successful design.

Several different electron multiplier configurations are in use. The design of the section between the photocathode and first dynode is quite important. It is necessary to arrange that all the photoelectrons from the large area of the photocathode are focused onto the relatively small area of the first dynode. Ideally to minimise subsequent transit time spread all photoelectrons would be focused to the same point on the first dynode. In practice, with careful focusing electrode design, a collection efficiency of about 90 per cent with a focused spot size of 5 to 10 per cent of the photocathode diameter can be achieved.

Figure 6.6. The squirrel-cage photomultiplier tube. This is the most compact dynode configuration available.

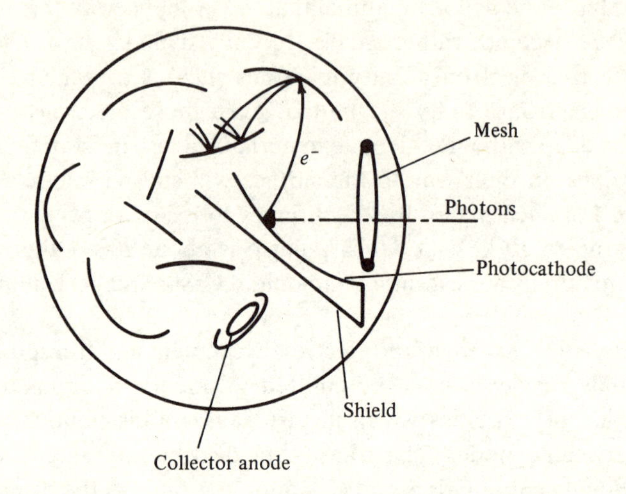

Two other common dynode configurations found in commercial photomultiplier tubes have properties which make them rather unsuitable for astronomical applications. The 'venetian-blind' configuration shown in Figure 6.7 gives a high electron multiplication in a small space and the large effective dynode area can be easily coupled to large (20 cm diameter) photocathodes. However the accelerating electric field between the dynodes is not very high with the consequence that the charge pulses reaching the anode are rather broad, and this severely limits the maximum photon counting rate possible. The large photocathode and first dynode also increase the tube dark count. The 'box-and-grid' configuration shown in Figure 6.8 gives efficient dynode-to-dynode electron collec-

Figure 6.7. The venetian-blind photomultiplier tube.

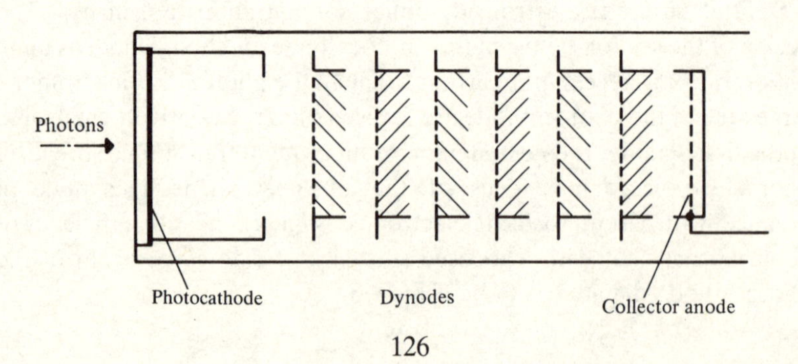

tion but again gives poor transit time dispersion and hence rather broad output charge pulses.

Figure 6.8. The box-and-grid photomultiplier tube.

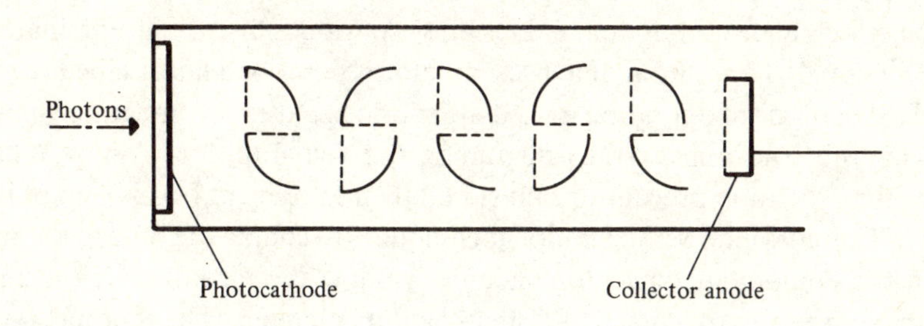

6.3 Practical photomultiplier tube operation

The principal reason for using photomultiplier tubes in astronomy is to achieve virtually noiseless photoelectron amplification within the tube so that the accuracy of a particular measurement is determined by the randomness of the light being measured, that is the photon 'noise' as defined in Chapter 1, rather than the external circuitry connected to the tube. The sources of noise present within the tube are therefore of fundamental importance in determining the low light level sensitivity. PMTs can be operated in either the pulse counting mode in which the individual charge pulses collected by the anode are amplified and counted with one count being equivalent to one photoelectron, or in the direct current (d.c.) mode in which the average number of charge pulses per second is detected as a current flowing through an output load resistor, and often shown as a galvanometer deflection on a strip chart recorder. In the absence of noise, both modes would be equivalent. In actual practice the pulse counting mode gives a better signal-to-noise ratio by allowing the effect of some noise sources to be suppressed.

The main source of noise in an uncooled PMT is thermal emission from the photocathode and first dynode. Since thermal electrons from the photocathode are indistinguishable from true photoelectrons and are produced even in the absence of any photocathode illumination, they are usually collectively called either the dark current or the dark count depending on the mode of operation in use. This dark count can be eliminated by cooling the whole tube, usually to below −20 °C. In recent years thermoelectric coolers have become commercially available so that operating the tube at a constant low temperature is no longer a problem. Thermal electrons from the first dynode undergo one less stage of amplification in reaching the anode and result in output charge pulses lower in amplitude than those from photoelectrons. Consequently the use of a pulse height discriminator enables these spurious pulses to be eliminated. The detailed operation of a discriminator is described later. However, if the d.c. mode of operation is used the spurious pulses are integrated along with the true pulses, thus degrading the signal-to-noise ratio. When cooling the tube precautions must be taken to prevent frosting up of the PMT input window. A common technique is to couple the tube to the rest of the equipment through a vacuum window the face of which is kept above freezing by means of a small heating element. The thermal emission from the photocathode is approximately proportional to the area of the photocathode so that a PMT should be selected with as small a photocathode as possible. The effect of the inevitable small variation in sensitivity over the photocathode can be minimised by designing the optical coupling into the PMT so that about 70 per cent of the cathode is illuminated by a defocused light spot.

At very low photon rates, spurious pulses can be seen which originate from cosmic rays and the decay of radioactive isotopes present in the PMT window, and these set the minimum photon rate detectable. It is found that exposure of a PMT to bright light such as daylight even with the electrode supply removed causes a large increase in dark count accompanied by a reduction in life. Recovery of the dark count to its previous low level takes about a day. It is strongly recommended that tubes intended for astronomical uses never be exposed to daylight.

The correct operation of a PMT requires the anode to be operated at a potential of one to two thousand volts positive with respect to the photocathode. Since external electronic measurement circuitry is usually connected to the anode, it is common practice to operate the photocathode at a high negative potential and the anode close to earth. The presence of this high voltage gives rise to a leakage current which flows along the outside of the tube and inevitably through the anode load resistor to ground. In the d.c. mode of operation this leakage current will be included in the total measured current whereas in the pulse counting mode it will not be seen at all.

The dynode secondary electron coefficient is quite a strong function of the dynode-to-dynode potential difference. For this reason the electrode power supply must be regulated to better than 0.1 per cent and be free from any mains ripple. It has been found that minimum output charge pulse width and minimum pulse width dispersion occur when the cathode to first dynode ($k-d_1$) voltage is as high as possible. This is often achieved by using a Zener diode regulator so that the $k-d_1$ voltage will not be affected by changes in the high voltage which are sometimes made to alter the overall PMT gain. A typical resistor chain used to derive the individual dynode voltages and connected to a ten-stage PMT is shown in Figure 6.9. The capacitors across d_7, d_8, d_9 and d_{10} are to supply the extra current needed when the amplified charge pulses encounter those dynodes. The load resistor, R_L, is chosen to match the impedance of the transmission line connected to the anode and is often 50 ohms.

The paths followed by the electrons within the PMT will be affected by any magnetic field. In general the electrons will be deflected away from their normal landing areas and the PMT gain will drop. This problem could be particularly severe in the case of an instrument mounted on a telescope since the relative orientation of the PMT and the terrestrial magnetic field will be changing throughout the night. High-mu magnetic shields are commercially available and often incorporated into the cooling unit so that the magnetic sensitivity of the PMT is in practice not a problem.

The pulse mode operation of a PMT is considerably enhanced, especially at low light levels, by the addition of a pulse height discriminator between the PMT and the electronic counting equipment. A discriminator has two important characteristics – hysteresis and dead time. Ideally an output pulse would be obtained for every input pulse whose amplitude exceeded some threshold value; no output would be obtained for input pulses below the threshold. In practice there is a region of uncertainty around the input threshold where an output may or may

Figure 6.9. A typical resistor divider chain connected to a ten-stage photomultiplier tube. The zener diode maintains the photocathode (k) to first dynode (d_1) potential constant. The capacitors are to provide dynode current pulses under high gain conditions.

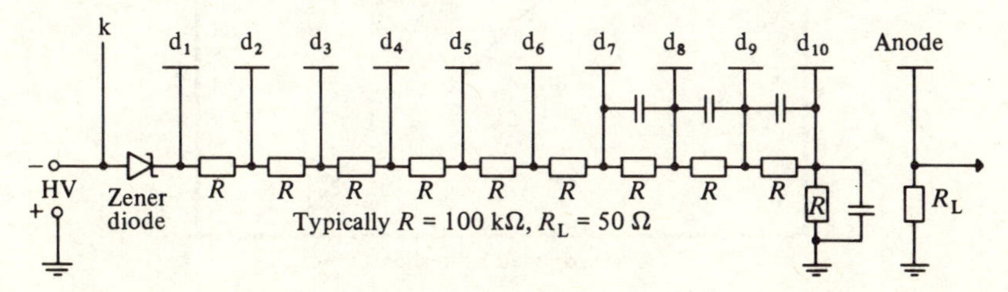

not be obtained. This region of uncertainty is known as the hysteresis and is typically about 1 per cent of the threshold. The dead time is the minimum time interval between two pulses larger than the input threshold for which two output pulses will be counted. In the ideal case this time would be zero but usually there is a small time of perhaps a few nanoseconds during which the discriminator is recovering from the previous pulse. It is important to know how the discriminator will handle pulses presented to the input during this time since this will affect the perceived counting rate on bright objects. A good discriminator will reject pulses received during the dead time so that a maximum count rate is clearly observed. It is often necessary to amplify the voltage signal developed across the PMT load resistor before applying it to the discriminator input. A typical PMT output voltage across a 50 ohm load would be 10 millivolts whereas most commercial discriminators require signals some ten times higher than this for correct operation. Of course, the amplifier must have sufficient bandwidth and a low enough noise level so that the voltage pulse is not degraded.

The discriminator is set up as follows. The PMT is connected to its high voltage supply and set up to view a weak steady light source viewed through a shutter so that as the shutter is opened and closed the output of the tube is alternately signal plus noise or noise alone. The discriminator input is connected to the tube via a suitable amplifier and a pulse counter is connected to the discriminator output. With the discriminator at its

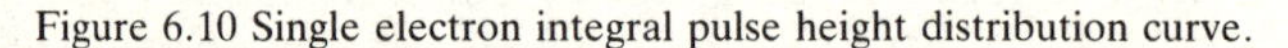

Figure 6.10 Single electron integral pulse height distribution curve.

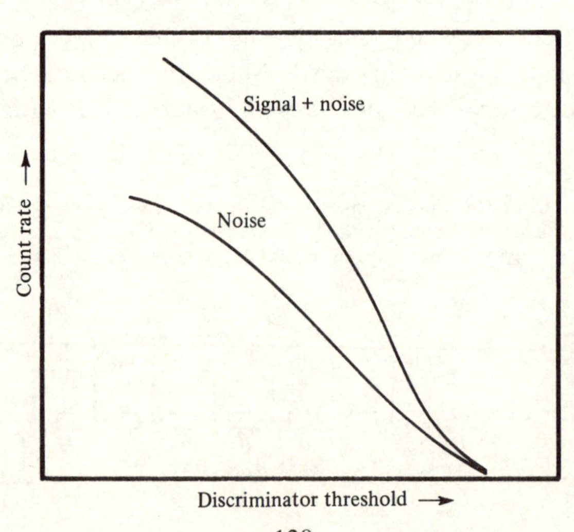

lowest threshold the count rate for both signal plus noise and noise is recorded. The discriminator threshold is then increased slightly and new count rates recorded. This continues until only a few signal counts are being recorded. A plot of total counts versus discriminator threshold is known as the single electron integral pulse height distribution curve and is similar to that shown in Figure 6.10. This shows that the optimum signal-to-noise figure occurs at low discriminator thresholds. Plotting only the incremental increase in count rate noted at each discriminator setting gives the differential distribution curve shown in Figure 6.11. Here it can be easily seen than the noise pulses are smaller than the signal pulses and an optimum setting for the discriminator would be at a threshold midway between the noise peak and the signal peak. The narrower the peak in count rate, the better the PMT for the pulse counting mode of operation. The shape of the peak is determined by the pulse spreading properties of the electron multiplier section of the PMT. An alternative but related way of establishing the optimum operating conditions is to set the discriminator threshold such that most signal pulses are passed and then to vary the PMT high voltage supply while monitoring the discriminator output count rate. If the PMT has good pulse counting properties, the count rate will initially increase rapidly with high voltage and then reach a pseudo plateau region before resuming its rapid rise, as

Figure 6.11. Single electron differential pulse height distribution curve.

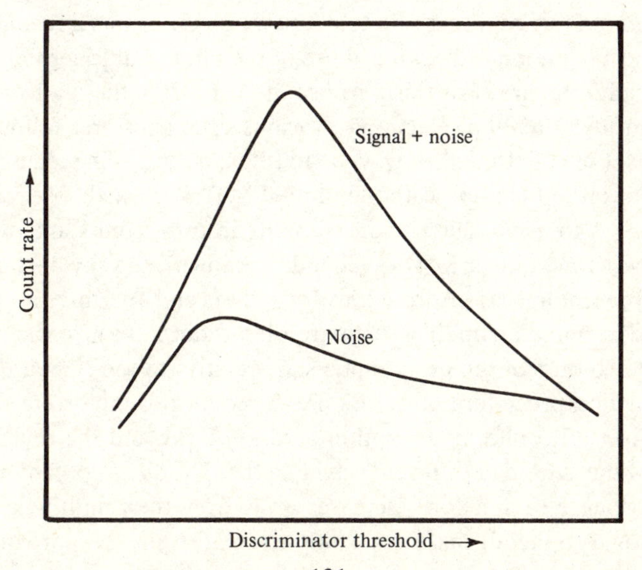

Figure 6.12. Typical count rate behaviour of a photomultiplier tube to variations in anode to photocathode high voltage.

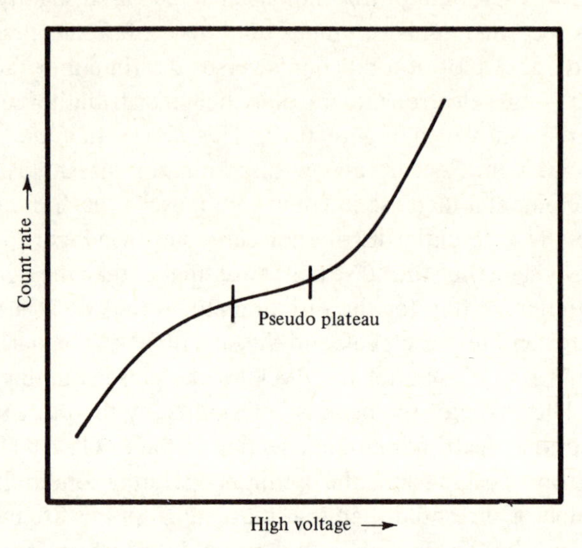

shown in Figure 6.12. The value of high voltage at the plateau is the optimum for pulse mode operation and corresponds to the narrowest charge pulse shape at the PMT anode. In all these experiments the count rate must never be so great that the PMT maximum anode current rating is exceeded. In fact for bright astronomical objects, this rating sets the maximum photon counting rate to about a million counts a second.

The most common astronomical use of the photomultiplier tube is for aperture photometry. For this purpose the light from the telescope first passes through one of a selection of various sized apertures followed by one of a set of coloured glass waveband filters before being focused by a Fabry lens on to the photocathode of the PMT. The usual observational procedure is to accumulate photon counts in turn from standard stars (whose brightness in various wavebands is known), the sky background and the programme stars through various filters and apertures according to a predetermined plan. The data are then reduced to give the magnitudes and colours of the programme stars relative to the standard stars. The reduction procedure must take into account two important effects: the varying atmospheric absorption across the sky and the relative response of the system to different wavelengths. The absorption of light by the atmosphere (extinction) increases away from the zenith by a factor proportional to the air mass in the line-of-sight, which is approximately proportional to the secant of the angle between the zenith and the telescope axis. In order to obtain a direct measurement of the pattern of absorption over the sky, and to obtain a zero-point for the total absorption, a number of standard stars, whose true magnitudes are accurately known, is observed periodically throughout the night. The relative wavelength response of the system, known as the colour equation, is usually very stable and need be checked only once per observing run unless one of the system components, in particular the photomultiplier tube, is changed. The colour equation is derived from the difference in observed and known magnitudes of a number of standard stars observed over as wide a colour range as possible. This technique allows a subsidiary set of standard stars to be set up covering a specific magnitude range in a certain region of the sky for the purpose of calibrating a photograph of the same region. A slightly different photometric technique is differential photometry. This involves either repeated observations of a known standard star and the programme star or the use of a two-channel instrument for simultaneous observations of the two stars. For accurate photometry, the atmospheric scintillation should be low and the extinction must be constant throughout the night.

6.4 Channel electron multipliers

The concept of a multi-dynode electron multiplier has been extended to its logical extreme of a continuous distributed dynode in the channel electron multiplier. This device, shown in Figure 6.13, consists of a hollow tube made out of semiconducting glass with a typical diameter of a few millimetres and a length of 100 millimetres. The inner surface of the tube, the channel, acts as both the photocathode for incoming short wavelength photons and the continuous distributed dynode. The channel has a typical resistance of 10^9 ohms. A positive voltage of several kilovolts applied between the ends of the tube establishes an accelerating electric field in the channel. Photons or electrons which strike the channel wall with sufficient energy release electrons which are accelerated along the channel axis while drifting across to strike the channel wall and release further secondary electrons. This process is repeated throughout the remainder of the channel until at the anode pulses of about 10^8 electrons are collected. The output pulse distribution is modified by space-charge effects and has the truncated non-exponential form shown in Figure 6.14. The width of the gain distribution is typically 50 per cent of the mean gain.

Figure 6.13. Channel electron multiplier. Photons strike the open end of the tube and release electrons which are accelerated towards the anode while drifting across the tube to strike the walls releasing secondary electrons.

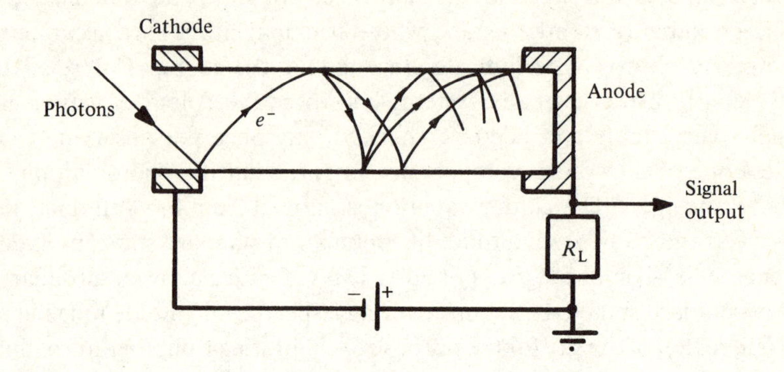

Figure 6.14. Output pulse height distribution curve for a typical channel electron multiplier. Typically the mean gain is 10^8 and the full width of the distribution curve at half maximum is 30 to 60 per cent of the mean gain.

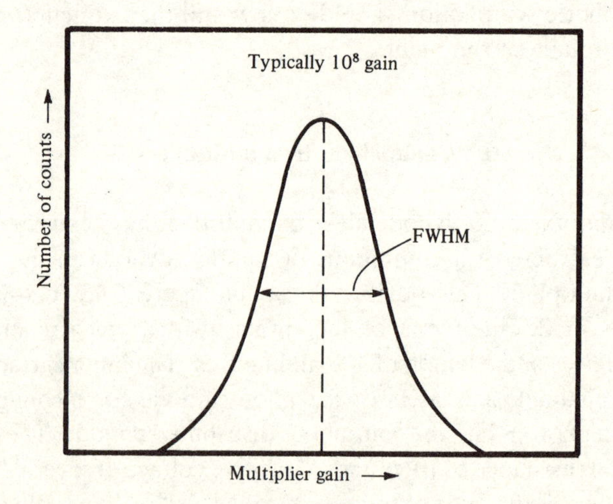

The channel is usually curved to prevent positive ions, which are accelerated in the opposite direction to electrons, reaching the front end of the channel where they could release electrons which would produce spurious output pulses. The work function of the semiconducting glass is so high that a separate photocathode must be employed for visible light wavelengths. This can be either an opaque layer deposited on the inside of the front end of the channel or a separate semitransparent photocathode such as used in photomultiplier tubes and which is electrostatically coupled to the channel input.

Since the parameter which most strongly defines the channel performance is the ratio of length to diameter of the channel, fibre optic technology has been used to reduce the channel diameter to the 20 to 50 micron range. Large numbers of these microchannels can then be bundled together to form a microchannel plate (MCP) which is used as either an imaging detector or, if all the anodes are connected together, a large area detector. It is also possible to use a phosphor screen at the output so that the whole MCP acts as an image intensifier. The earlier MCPs employed straight channels and suffered from positive ion feedback problems at high gains and pressures greater than one microtorr. This problem has been partially overcome by cascading two plates in the 'chevron' configuration shown in Figure 6.15. More recently it has become technically feasible to employ curved microchannels in an array. This enables a single MCP to be operated at stable gains of 10^6 without the ion feedback problems and short operational lifetimes of the chevron plates. When coupled to multi-element anode detector arrays, the MCP makes a very compact two-dimensional photon counting detector particularly suitable for space-borne applications.

Figure 6.15. Two cascaded microchannel plates arranged in a chevron configuration. Electrons pass from the first MCP to the second MCP whereas the positive ions cannot travel back to the entrance of the first MCP.

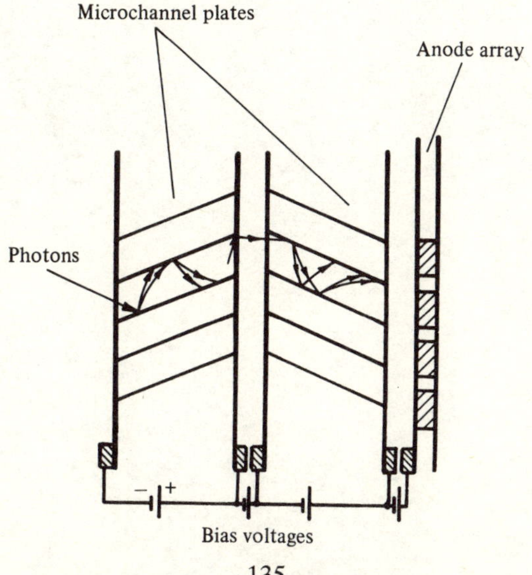

References for Chapter 6

EMI Photomultipliers Manual, EMI, 1979.
Methods of Experimental Physics, Vol. 12A, Academic Press, 1974.
RCA Photomultiplier Manual, RCA, 1970.

7

Television-type imaging detectors

The most important difference between single channel and imaging detectors is the need for a storage mechanism to hold an electrical analogue of the optical image since the majority of imaging detectors allow for the readout of only one small part of the image at a time. It is possible to construct an imaging detector from individual elements such that all elements can be read simultaneously. However, the large number of resolution elements required for typical imaging applications combined with the close proximity of elements necessary to minimise dead spaces between the elements essentially precludes the design of such a parallel output detector. Instead, virtually all successful imaging detectors operate by sequentially reading the resolution elements and sending the data to an external storage medium where the original image can be reconstructed later. The use of a storage medium leads to a set of considerations not unlike those of the photographic plate, for example resolution, dynamic range, linearity and others already mentioned in Chapter 2. The development of vacuum tube imaging detectors has followed closely the demands for ever better and better television camera tubes by the broadcasting community. For this reason vacuum tube imaging detectors are often referred to as television-type imaging detectors. The common characteristic of all television-type camera tubes is the use of an electron beam to read out the video information which is stored as a charge pattern on an element called the target.

7.1 Electron beam target readout

The target of a camera tube contains the electronic charge analogue of the object being viewed in the sense that the charge density at a particular point on the target is proportional to the number of photons that have fallen on the corresponding point of the photocathode since the target was last read. Figure 7.1 illustrates the basic operation of target storage and readout for a typical tube used in modern astronomical detectors. Photons incident on the tube faceplate are carried by a fibre optic bundle to the curved photocathode where they release photoelectrons with a typical quantum efficiency of 20 per cent. The photoelectrons are accelerated through several kilovolts and strike the target. The target is a semiconductor in which the photoelectrons each produce many electron–hole pairs by the same mechanism already described for the dynodes used in photomultiplier tubes. In the case of most targets, however, the photoelectrons are of higher energy and thus produce the electron–hole pairs deep within the target. Secondary electron emission is therefore negligible. Prior to illumination of the photocathode an electron beam is scanned over the rear of the target. The electron beam is virtually a perfect conductor so that the rear of the target assumes the potential of the source of the beam, that is the cathode, which is normally at earth potential. The front of the target is held at a fixed positive bias potential of a few tens of volts. The target is usually made of a resistive material and can be thought of as a leaky capacitor charged up to the target bias voltage. In the absence of the electron read beam, electron–hole pairs produced in the target will locally discharge this capacitor, raising the potential of the rear of the target towards the fixed-bias voltage. Those areas of the target receiving the greatest photoelectron flux will have the most positive rearside potential. When the rear of the target is again scanned with the electron read beam, the most positive or 'brightest' areas of the target will require the highest number of electrons to restore the potential to earth. This recharging electron current flowing in the target is then the video output signal. The output signal current is often 1 nA or less and requires amplification at the camera before being sent to the main signal processing and storage electronics. It is good engineering practice to mount the amplifier as close

Figure 7.1. Basic operation of target storage and readout in a direct beam camera tube. G1 is the control grid, G2 is a cylindrical lens, G3 is an accelerating wall electrode and G4 is the decelerating mesh.

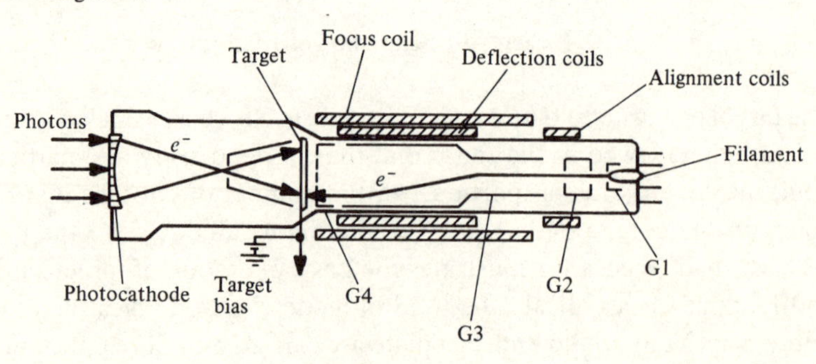

to the target as possible and to cool it along with the camera tube. The signal processing usually consists of further amplification, charge integration to reduce random noise and analogue to digital conversion.

While the quantum efficiency of the tube is defined by the photocathode, linearity and image latency or lag are determined by the target. The linearity of the tube which is usually defined as the slope of the log output current versus log input flux curve, more concisely referred to as the gamma, must be as near unity as possible for astronomical applications. However, as will be seen shortly, this is not usually the case in television broadcasting, which limits the choice of suitable tubes. Image latency or lag is due to the resistive nature of the target, which means that recharging takes a finite time. Under usual scanning speeds a small fraction of the positive charge distribution will remain after one and often more scans. The image usually decays exponentially and the image lag figure quoted for a particular camera tube refers to the percentage of image charge on the third scan after illumination has been removed.

Figure 7.1 also shows the typical arrangement of electrodes and coils used to generate, focus and deflect the electron read beam. Electrons from the indirectly heated cathode pass through the first control grid, G1, whose aperture defines the minimum beam diameter and whose potential with respect to the cathode defines the beam intensity, through a cylindrical lens, G2, accelerating electrode, G3, and finally a deceleration mesh, G4, which ensures that the beam lands everywhere orthogonal to the target, an important feature in reducing the variation of sensitivity across the target. Since a magnetic focusing lens has much lower spherical aberration than an electrostatic lens, the major part of the beam focusing is provided by a medium length solenoid spanning the total electron beam path. The longitudinal magnetic field acts on the radial velocity component of diverging electrons leaving the cathode and directs them back towards the central axis. In this way an image of the first grid aperture, G1, is formed at the target. If however, the electron beam enters the focusing field off axis, the beam focused at the target will not be circular but distorted. To eliminate this effect vertical and horizontal alignment coils placed before the focus coil are used to give a weak beam centering field. Horizontal and vertical deflection coils mounted orthogonally inside the focus coil generate the transverse magnetic fields for scanning the target. For the high beam stability demanded by astronomical applications, the usual electrode and coil power supplies found in commercial cameras are rarely stable enough. Specially designed power supplies stable to at least 0.1 per cent are usually required.

The dynamic range and sensitivity of a tube are set by the relative magnitudes of maximum permitted target current and target leakage current in the absence of illumination, usually called dark current. In manufacturers' data sheets these currents are specified in nanoamps at television scan rates whereas the astronomer is more interested in the maximum charge that the target can store and the rate at which thermal electron–hole pairs are generated in the target. An exact equivalence can be calculated only if the pixel size, target size, which is usually equal to the photocathode size, and scanning speed are known. However it is possible to derive typical figures based on pixel sizes and slow scan rates often used in astronomy. Taking a pixel size of $40 \times 40\ \mu m$, a pixel readout time of 50 microseconds and a target diagonal of 16 mm, a target current of 100 nA is equivalent to about 10^7 stored charges per pixel per frame. The total number of pixels in the frame would be about 80 000 and the time required to read one complete frame would be four seconds. This is to be compared with a typical TV frame readout time of 33 milliseconds. A dark current of 1 nA is equivalent to a thermal electron–hole pair production rate of 3000 per pixel per second. These figures can be easily scaled to fit actual tubes and different astronomical slow scan modes.

The actual electrode and target configurations used in the most common types of camera tube differ quite considerably from tube to tube. This in turn affects the suitability of a particular tube for use as an astronomical detector. These various types of camera tube will now be described in some detail.

7.2 Return-beam tubes

The earliest camera tubes having sufficient sensitivity to be used on a routine basis for television work were the return-beam tubes, the image orthicon and the image isocon. While this class of tube has been largely superseded by more sensitive tubes, the return-beam tubes have been used to make astronomical observations, and deserve a brief description.

Return-beam camera tubes are different from the example described in the previous section in that the electron optics are so arranged that the fraction of the electron read beam not required to charge the target back to earth potential is directed back to the vicinity of the electron gun where the video information present in the beam can be extracted. Figure 7.2

Figure 7.2. Principal components of the image orthicon return-beam camera tube.

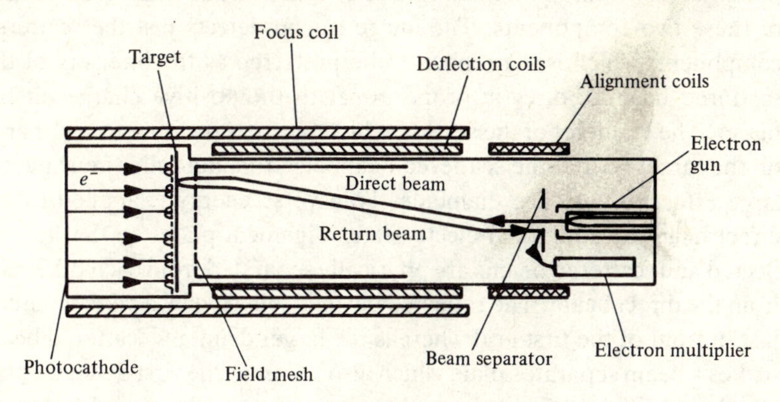

illustrates the image orthicon camera tube which was developed in the 1940s. Electrons emitted from the photocathode are accelerated through a few hundred volts onto the magnesium oxide target where about 15 secondary electrons are released per primary photoelectron. Owing to the low energy of the photoelectrons and the nature of the target material, the performance of the target is closer to that of the first dynode of a photomultiplier tube rather than a typical target. Secondary electrons are released from the front surface of the target and collected by a fine mesh operated at a few volts positive with respect to the target. This leaves behind a positive charge distribution on the target which is an analogue of the light distribution on the photocathode. The scanning beam deposits electrons on the target until the positive charge is neutralised. The scanning beam is then reflected back towards the cathode where it strikes the first dynode of an electron multiplier which has sufficient gain to remove the need for a very sensitive amplifier. Since the video information is the difference in intensity between the direct beam and the return beam, the low light level sensitivity is set by the thermal noise or statistical energy variations in the electron beam. This noise is accentuated at very low light levels where the thermal energy spread can be larger than the differences in charge on the target. One way to increase sensitivity is to increase the image section gain. This was not technically possible in the 1940s when these tubes were under development. Another way is to change the manner of extracting the video information from the return beam. This was adopted in the image isocon.

The return beam actually contains two components, a specularly reflected beam and a scattered component due to interaction with the potential distribution on the target. The image orthicon detects the sum of these two components. The image isocon detects just the scattered component, which is very much to be preferred as the intensity of the scattered beam is directly proportional to the positive charge on the target. The diameter of the scattered beam increases with positive charge on the target so that the scattered beam electron multiplier must have a large effective entrance diameter. Transverse energy is applied to the direct beam by additional electrostatic alignment plates so that the reflected and scattered beams are physically separated from each other and from the direct beam. The reflected beam is collected by a plate mounted just in front of the first grid whereas the larger diameter scattered beam strikes a beam separator plate which also serves as the first dynode of the electron multiplier. This electrode arrangement is illustrated in Figure 7.3.

Despite the more complex electrode structure with its attendant greater mechanical and electrical tolerance requirements, the isocon can be routinely operated with both sensitivity and dynamic range some ten times higher than the image orthicon. However, despite this improvement in readout efficiency the image section gain is too low to give photon noise dominated performance without the use of external image intensifiers. The greatly increased image section gain of modern camera tubes combined with the availability of very low noise video amplifiers obviates

Figure 7.3. Return-beam separating electrodes used in the image isocon camera tube.

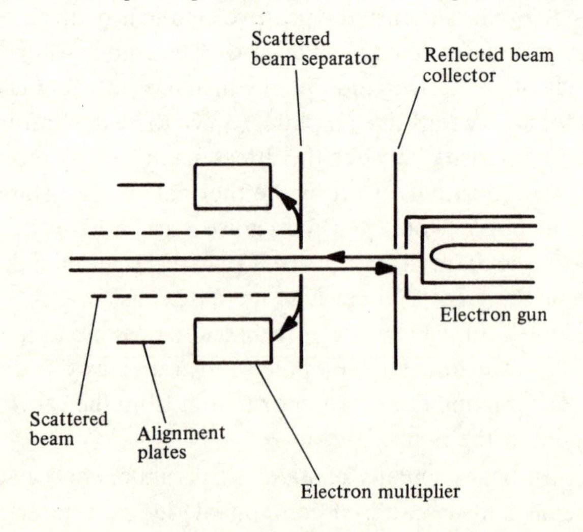

the need for the internal electron multiplier gain of return-beam tubes so that modern detectors employ direct-beam tubes.

7.3 Secondary electron conduction tubes

The term secondary electron conduction (SEC) can strictly be applied to any camera tube in which primary photoelectrons are accelerated towards a target where they produce secondary electrons which in turn modify the charge distribution within the target. However the term has generally come to mean a particular type of direct-beam tube having a target composed of low density potassium chloride. Figure 7.4 shows the typical SEC target in which an aluminium oxide layer about 0.7 μm thick supports an aluminium signal plate of similar thickness on to which a 20 μm thick layer of low density potassium chloride is evaporated. In normal operation the rear of the target is taken to earth potential by scanning with an electron beam prior to illuminating the photocathode. As the signal plate is kept at a typical positive potential of about 30 volts and the target conductivity is very low compared to the aluminium an electric field will be established within the KCl. Photoelectrons with typical maximum energies of 10 keV pass rapidly through the relatively thin $Al_2 0_3$ and Al layers and dissipate their energy in the KCl by creating secondary electrons. As about 30 eV is required to create one secondary electron, a maximum target gain of about 300 can be achieved – a factor of 20 better than the return-beam tubes. The electric field in the KCl moves the electrons towards the signal plate leaving behind a positive charge on the rear of the target. If the photoelectron flux is sufficiently

Figure 7.4. The internal layout of a typical secondary electron conduction (SEC) camera tube target.

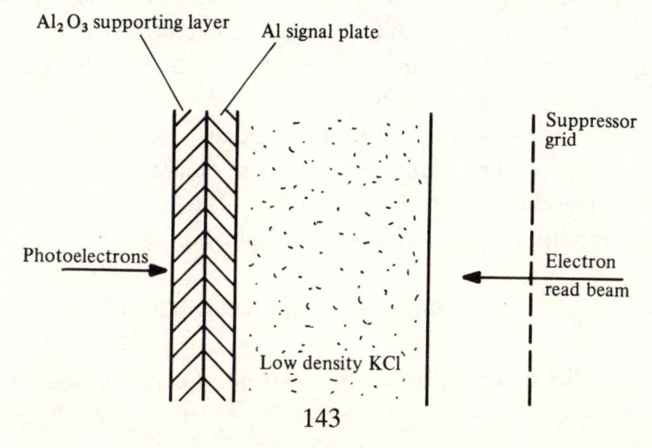

great the rear of the target can approach the signal plate potential and cause a drastic reduction in the electric field within the KCl. This increases the probability that the newly produced electrons and holes will recombine before being swept apart. This non-linearity is extremely undesirable in an astronomical detector but can be prevented by either reducing the target integration time or reducing the high voltage in the image section of the tube. The latter solution is usually more desirable as it corresponds to an increase in photon flux for a given target readout current. It is even possible for the rear of the target to become more positive than the signal plate whereupon the electric field is reversed and secondary electrons are ejected out of the rear of the target.

The discharging of the target by the electron read beam causes a current to flow through the signal plate directly proportional to the positive charge distribution on the target rear and constitutes the video signal as in all direct-beam tubes. A negatively biased suppressor grid is mounted close to the rear of the target to prevent secondary electron ejection by the read beam which can happen if the target rear becomes too positive. For broadcast television applications the target is read continuously so that each pixel integrates charge for much less than one second. However, for astronomical applications the low photon fluxes encountered make pixel integration times of hours desirable. This can be achieved by simply switching off the electron read beam for the desired integration time, during which time the target acts solely as a storage medium. However, this is only feasible if the thermal electron–hole production rate, usually called the dark current, is negligible. The great advantage of the KCl target is that the dark current at room temperature is at least 100 times lower than for any other target. In fact it is low enough to permit hour long integrations while running uncooled. Other camera tubes require cooling to −50 °C or lower to achieve the same results. As a result these tubes are found at many large observatories and if coupled to an image intensifier tube they can detect individual photons. Maximum target current is about 300 nA.

Image lag is as low as any other tube, being 3 per cent of signal in the third frame. Typical resolution in the centre of a 25 mm diameter photocathode at a modulation transfer efficiency of 50 per cent is 25 μm – quite adequate for the majority of astronomical purposes.

7.4 Vidicon and lead oxide vidicon

The term vidicon generally refers to a class of camera tubes in which a photoconductive target serves as both imaging medium and charge storage medium. However, as in the case of the SEC tube, the term has come to mean a specific type of tube, namely a direct-beam tube having an antimony trisulphide target. Figure 7.5 illustrates the basic operation of the vidicon. Photons enter the photoconductive Sb_2S_3 target through a transparent faceplate electrode and if sufficiently energetic, that is having a wavelength shorter than 650 nm, produce electron–hole pairs in the target. Provided the transit time through the target is greater than the lifetime of the charge carriers, the electrons will be swept towards the faceplate electrode leaving behind a positive charge on the rear target face. The establishment of an electric field within the target and the reading of the target by a low energy electron beam are similar to that in the SEC tube.

The use of the one target to perform two separate functions leads to two operational characteristics of considerable disadvantage for astronomical purposes: severe image lag at low light levels and a gamma of less than unity. The electron read beam when discharging the target actually deposits electrons until the target is slightly negative with respect to the cathode. There is no gain mechanism operating in the target so that the change in target potential for very low light levels will be small and may not even be sufficient to make the target positive with respect to the cathode. As a result, when the target is read again the electron beam sees a retarding electric field under which conditions a great number (perhaps ten or more) of read scans will be required to discharge the target fully. This low light level image lag problem is common to all classes of vidicon and can be greatly reduced by temporarily increasing the target bias

Figure 7.5. Essential components of the direct-beam vidicon camera tube.

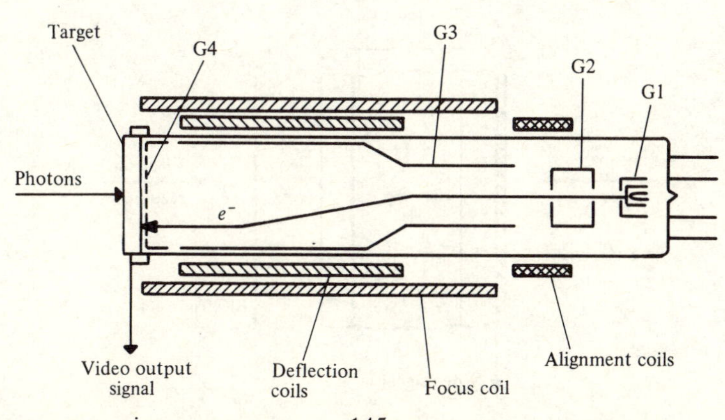

voltage immediately prior to reading so that the video information sits on a pedestal of typically one volt. This is easily removed later in the signal processing chain. This image lag is further worsened in the Sb_2S_3 vidicon by a tendency for some ions to become trapped within the photoconductor. Charge neutralisation then takes longer than would be expected from a simple consideration of the bulk target resistivity. The gamma of less than unity is thought to be a consequence of the small but significant probability of charge pairs recombining before being swept apart by the target electric field.

The Sb_2S_3 vidicon also suffers from lateral spreading of the positive charge pattern which limits the resolution under long integration times and causes blooming, that is spreading of bright points in the image, under high contrast conditions.

To eliminate many of these drawbacks the lead oxide vidicon, also known by the trade names Vistacon and Plumbicon, has been developed. The many improvements in performance come from the replacement of the antimony trisulphide target with one of lead oxide, PbO, in the form of a p-i-n diode, as shown in Figure 7.6. The remainder of the camera tube is similar to the standard vidicon shown in Figure 7.5.

Under normal operation the front face of the target is operated at 50 volts positive with respect to earth. The rear of the target is set to zero potential by the electron read beam in the usual manner. The p-i-n diode is therefore reverse biased with nearly the whole 50 volts appearing across the intrinsic layer. The band gap energy of PbO is 1.9 eV so that photons of shorter wavelength than 650 nm will produce electron–hole pairs

Figure 7.6. The target structure of a lead oxide vidicon.

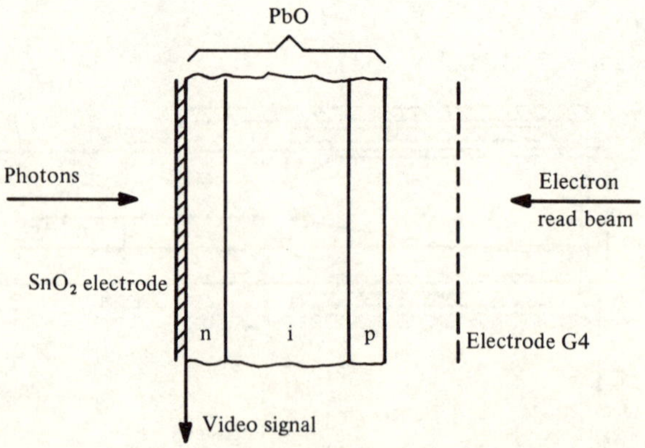

within the intrinsic region. The charge pairs are then rapidly swept apart by the electric field. This detection process is highly efficient and quite linear showing a typical quantum efficiency of 50 per cent at 500 nm and a gamma of 0.95. The actual spectral response of the lead oxide vidicon is shown in Figure 7.7 compared with other astronomically important camera tubes. The short-wavelength cutoff is caused by absorption of photons before reaching the intrinsic region of the target. The rather high value of band gap energy lowers the room temperature dark current to

Figure 7.7. Spectral response curves for four astronomically important camera tubes.

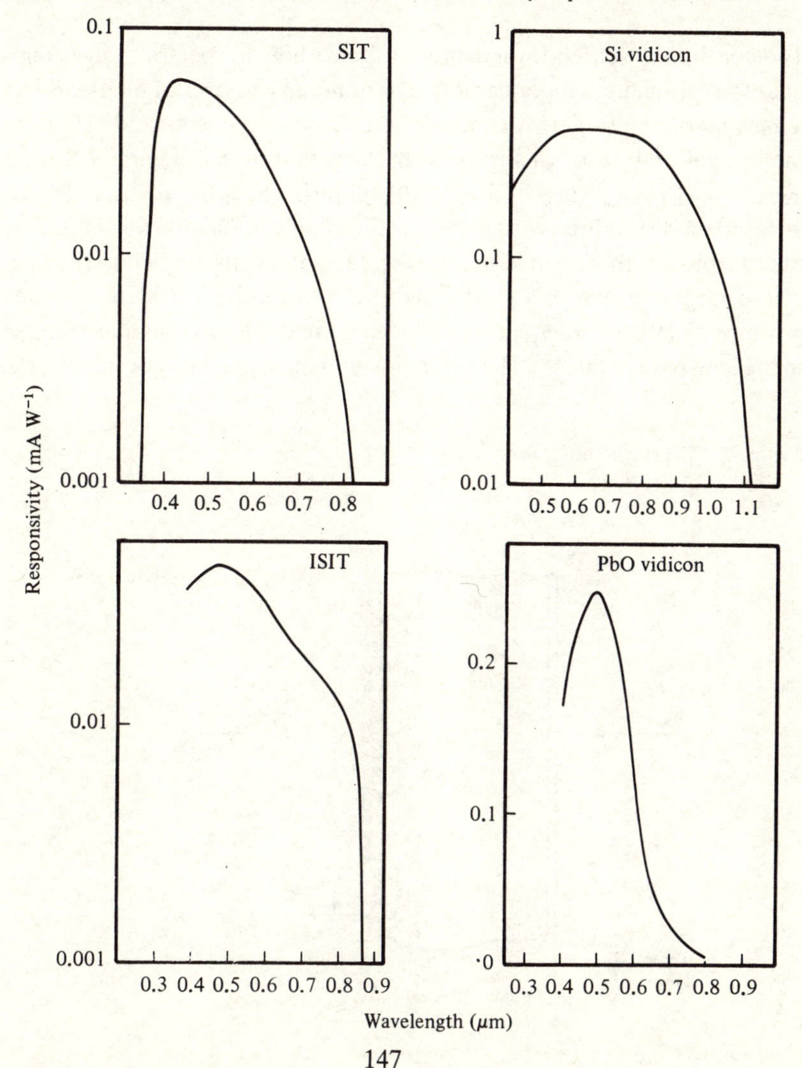

about 0.5 nA as compared to the 10 nA typical for the Sb_2S_3 vidicon. Of course the dark current can be reduced substantially by cooling the target.

The traps in the photoconductive target which cause the long image lag in the vidicon play only a minor role in the lead oxide vidicon so that the target recharging time is mainly determined by the combination of target capacitance and read beam current. The image lag is 3 per cent, the same as for the SEC tube. The typical spatial resolution is also the same as for the SEC tube at 25 μm.

7.5 Silicon target vidicon and SIT

The continuing search for a camera tube with wide spectral range, high quantum efficiency, unity gamma and immunity to optical overload has led to the silicon target vidicon. As in the case of the lead oxide vidicon, the special feature of this tube is the target, shown in Figure 7.8. The target consists of a slice of n-type silicon into which p-type 'islands' are diffused at 20 μm intervals. A thin layer of silicon dioxide insulates one island from another. A positive bias voltage of 8 volts applied to the face of the target reverse biases the array of p-n diodes. Incident photons produce electron–hole pairs, discharging the diodes in a similar manner to the lead oxide vidicon. The rather lower band gap energy of silicon (1.2

Figure 7.8. The silicon diode array target.

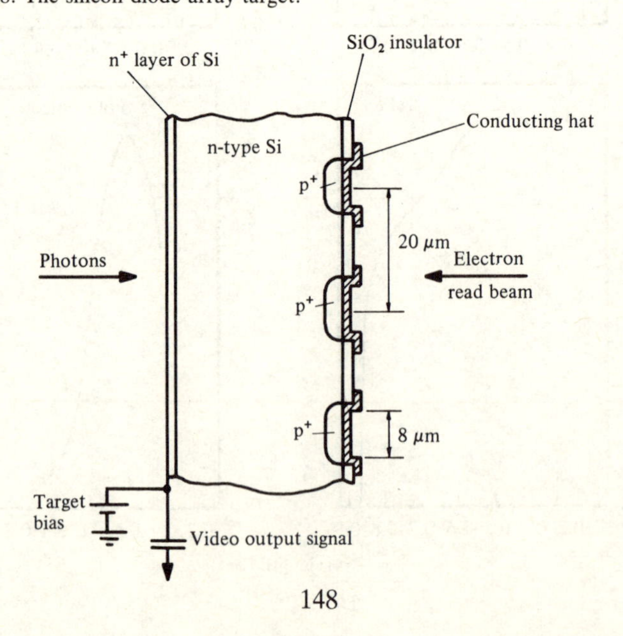

eV) as compared to PbO extends the red response to 1100 nm. The quantum efficiency remains at least 50 per cent over the wavelength range 400–800 nm. This considerable increase in spectral sensitivity is clearly shown in Figure 7.7.

The penalty to be paid for the improved red response is a ten-fold increase in dark current over the lead oxide vidicon. In astronomical applications, the silicon target vidicon must be cooled to below 0 °C otherwise the dark current will completely saturate the target in a matter of seconds. Cooling the target to dry ice temperatures (−78 °C) reduces the dark current to about 2 pA or 6 electrons per pixel per second. Operated in this mode, hour long integrations are possible before the thermally produced charge reaches 1 per cent of the target storage capacity. The p-n diode structure has been found experimentally to have a gamma equal to unity up to an integrated photon flux of 10^6 photons per pixel. Above this number the electric field within the target is reduced to the point that newly created electron–hole pairs have a significant probability of recombining before being swept apart.

The heavier doped n^+ layer on the front face of the target is to establish an electric field at this face which is strong enough to separate quickly those charge pairs produced by the shorter wavelength photons which do not penetrate very far into the target. Figure 7.8 also shows a conducting 'hat' covering the p-type islands and extending over part of the neighbouring SiO_2 insulator. This is to prevent a buildup of negative charge on the insulator which would eventually repel the read beam. An alternative approach is to coat the rear of the target completely with a thin conductive layer. However, although this prevents a negative charge buildup, it increases image lag and decreases spatial resolution. The image lag for the conducting hat case is three times worse than for the lead oxide vidicon on account of the increased capacity of the p-n junction as compared to the p-i-n junction. There is a significant increase in image lag at low light levels as mentioned earlier. An alternative approach to increasing the target bias potential prior to reading is to illuminate the target with a faint but non-varying source of light.

The silicon diode array structure has been found to be very stable with time and to be virtually immune to optical overload, so that the camera tube can safely be operated in full daylight. The diode structure does limit the maximum spatial resolution to one diode spacing or 20 μm. If the read beam is only slightly larger than this interference effects can produce a spurious modulation of the video signal. Hence a slightly defocused beam of some 30 to 40 μm diameter is often employed.

The less than unity target gain of all unintensified vidicons places great demands on the first video amplifier in detecting currents equivalent to a few hundred electrons per pixel. Conventional current-to-voltage amplifiers using special low noise field effect transistors have been designed with an equivalent input noise of 800 electrons per readout. More recently, charge sensitive integrating amplifiers have been used and found to have a much lower noise of 200 electrons. As the typical pixel storage capacity is equivalent to 10^6 electrons, the dynamic range of the system comfortably exceeds 1000 to 1.

Whilst cooled silicon vidicons have been used successfully for long term integrations of faint objects, the lack of target gain severely limits the range of objects which can be detected in only a few minutes integration. The inclusion of an image intensifier section in front of the target has greatly extended the use of the silicon vidicon by permitting the detection of single photons. This combination is known as a silicon intensified target vidicon or SIT, and is shown in Figure 7.9. Photons striking the photocathode, which is usually of type S20, release electrons which are accelerated through 10 kV in the electrostatically focused image section and strike the silicon diode array target. As only about 3.5 eV is required to produce one electron-pair, the target has a typical gain of 2000, some ten times higher than the SEC target. For brighter objects the gain is

Figure 7.9. The basic arrangement of the photocathode and silicon target in the SIT camera tube.

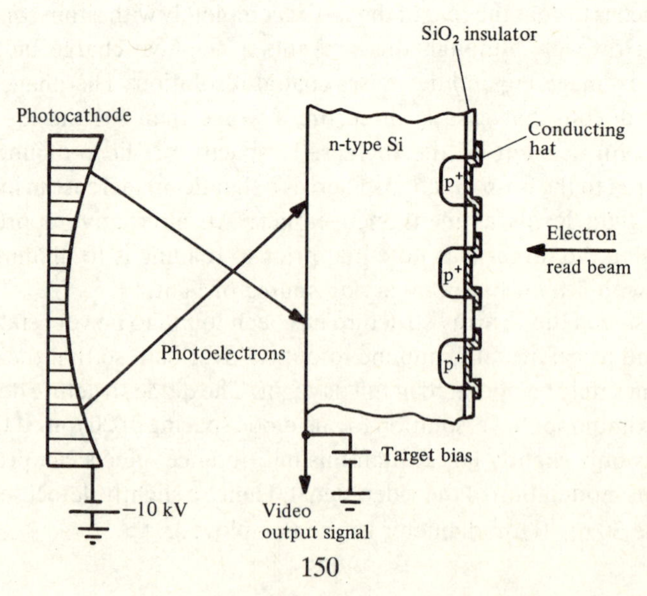

reduced by decreasing the high voltage in the image section. Unfortunately the red sensitivity and high quantum efficiency of the unintensified silicon vidicon are lost.

The maximum spatial resolution of the SIT is still set by the diode structure with 30–40 micron pixels being typical. With a highly focused beam, Moiré fringes have been seen and are caused by beating between the regular hole spacing in the deceleration mesh and diode structure in the target.

7.6 Intensified camera tubes

Intensification here means the addition of one or more stages of optical amplification before a particular camera tube so as to increase its sensitivity greatly by reducing the combined effects of readout noise and dark current. No camera tube available today can detect single photons while operating at room temperature. The SEC tube has very low dark current but insufficient target gain to overcome readout noise, the SIT tube has high target gain but a high dark current whilst the remaining types of tubes have both low gain and fairly high dark current. This can be a serious drawback in certain astronomical applications of a routine nature where the provision of cooling would create operational difficulties.

Electrostatically focused or magnetically focused image intensifiers are both used coupled to the following camera tube with lenses or fibre optics. Popular examples of the latter are the fully encapsulated electrostatically focused commercial ISIT and ISEC tubes, where the first I stands for intensified, often found in guiding and acquisition systems on the larger telescopes. A typical ISIT/ISEC is illustrated in Figure 7.10. Photons incident on the fibre optic input window release electrons from the photocathode deposited on the rear of the window. These electrons are accelerated through typically 15 kV and strike a phosphor screen deposited on the front of the output window. Approximately 200 photons are released by the phosphor for every incident electron which taking into account the typical photocathode quantum efficiency of 20 per cent gives an overall intensifier gain of 40. In an encapsulated tube the other side of the fibre optic output window is coated with the photocathode of the following SIT or SEC tube. For optimum transfer efficiency the spectral responses of the phosphor and photocathode should be similar. The total gain of the intensified camera can be as high as 80 000 for an ISIT and 3200 for an ISEC.

Another common approach is to lens couple a three-stage electrostatically focused image tube to a standard camera. The internal construction

Figure 7.10. The basic arrangement of electrodes in an ISIT or ISEC camera tube.

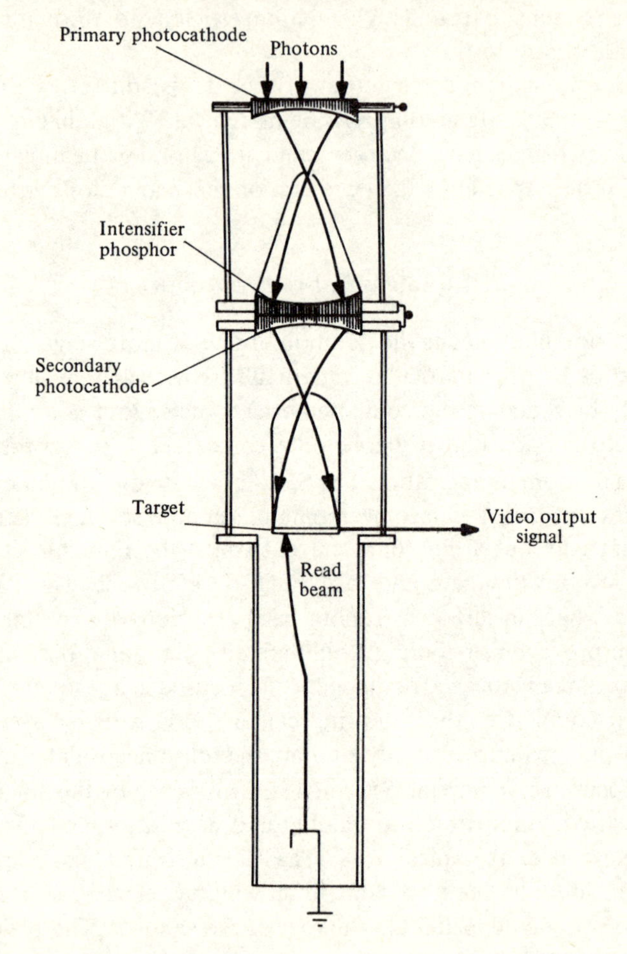

of a typical three-stage image tube is shown in Figure 7.11. The gain of a three-stage tube is at least 50 000 so that even though the coupling lens will have an efficiency of only a few per cent the combination of image tube and camera will still have high sensitivity. The high voltage (~ 45 kV) across the image tube can cause corona and high voltage leakage unless special care is taken in potting all exposed high voltage points, usually with silicone RTV potting compound.

Electrostatically focused image tubes suffer from distortion and loss of resolution at the target edges. Magnetically focused tubes give much less distortion and are used extensively. Possibly the best known system of

Figure 7.11. A typical three-stage electrostatically focused image tube.

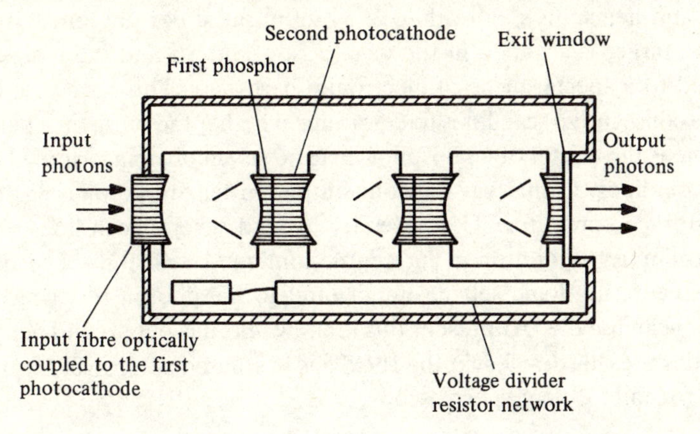

this kind is the University College, London, image photon counting system (IPCS). Figure 7.12 shows the four-stage magnetically focused image tube with an optical gain of 10^7 coupled by a 1 per cent efficient lens system to a commercial lead oxide vidicon camera. Taking into account the quantum efficiency of the first photocathode this means about 7×10^5 electron–hole pairs generated in the vidicon target for each primary photoelectron. This size of charge pulse is sufficiently large that the camera does not require cooling, especially considering the fairly low dark current of this type of vidicon. The current flowing through the focusing coils generates heat which if not removed by a cooling jacket would cause electron emission from the first photocathode indistinguishable from photoelectric emission. The size of the image produced on the

Figure 7.12. Sectional view of the image intensifier, coupling lens and television camera head used in the IPCS.

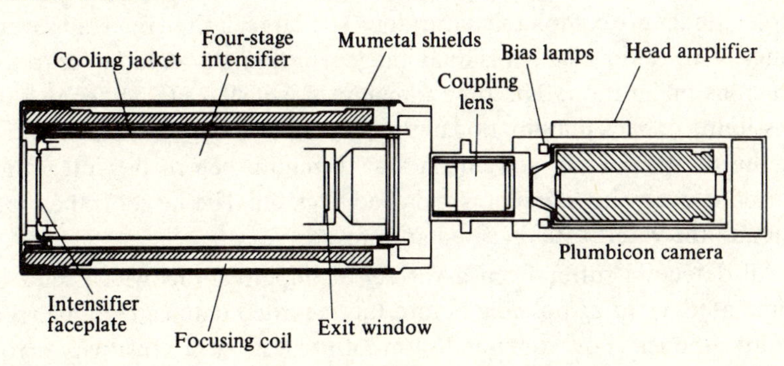

final phosphor by a single photon is large enough to cover several scan lines and hence give more than one output pulse per photon.

The target is scanned in the usual raster pattern and the video data passed to a special piece of electronics hardware. This hardware compares consecutive scan lines to determine whether the pulse on a particular line is the centre of a phosphor image or is simply image spread from adjacent lines. In this way only one output pulse per photon is passed to the storage computer. The necessity to read several scan lines before determining the photon event centre combined with the speed of the event centering logic sets an upper limit to the photon flux which the system can handle. At present this is a few photons per second per pixel but, despite this drawback, the IPCS has been used extensively, particularly for faint object spectroscopy.

7.7 Typical television-type detector operation

The most widely used camera tubes, in terms of published results, are the silicon intensified target vidicon, SIT, and the intensified Plumbicon used in the IPCS. Their major use has been to image the output of a spectrograph where the two dimensional nature of the detector enables the spectrum of the object being studied and the spectrum of the surrounding night sky to be recorded simultaneously.

Low resolution spectroscopy of faint objects and high resolution spectroscopy of bright objects results in low photon arrival rates so that hour long integration times are required. The IPCS handles this by continuously adding frames in the storage computer. The SIT employs target integration and requires additional operating steps. Firstly the camera tube must be cooled to below −50 °C. Secondly, the target must be prepared before integration can commence. The photocathode is exposed to a relatively bright diffuse light for a few seconds so that all target diodes are completely saturated. The target is then read out several times. This removes any residual image which might be left over from a previous integration. The tube filament is switched off, to remove the possibility of back illumination of the red sensitive target, and integration of the desired object started. About a minute before the integration period is over, the filament is switched back on. The target is then read out and the video data digitised and stored.

All detectors suffer from a variety of imperfections which must be eliminated as far as possible before the recorded data can be analysed. Photocathodes show a generally smooth variation in sensitivity across their surfaces. This effect is removed by taking a frame with the camera exposed to a diffuse light source such as the defocused Moon or the observatory dome lights and then dividing all data frames by this 'flat-field' frame. Following this step the night sky spectrum is subtracted to give the true object spectrum. It sometimes happens that the camera is not orientated correctly with respect to the spectrograph. Large computers are now available to rotate the image effectively by means of software. A reference spectrum from a standard spectral source such as an iron arc is used to calibrate and perhaps linearise the wavelength scale. Most targets have a few defective pixels which show up as bright spots on the recorded image. These can be removed by interpolating between adjacent pixels.

Of course, both the IPCS and the SIT as well as other two-dimensional

Figure 7.13. (a) Uncalibrated data from an exposure on the quasar Q2225-4127. The upper curve is the quasar + sky + dark offset signal. The middle curve is sky + dark offset signal. The lower curve is the difference of the upper two. (b) Flat-fielded and calibrated plot of Q2225-4127.

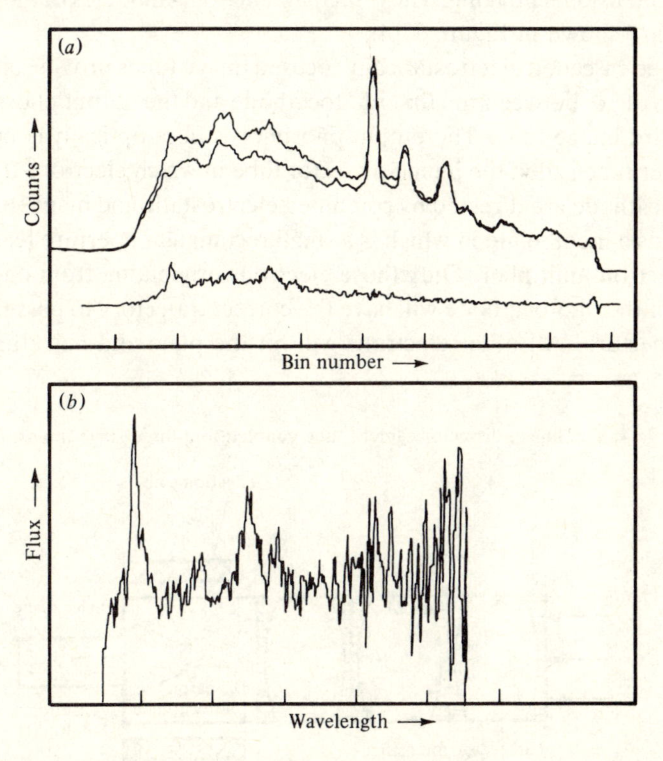

detectors have been used for direct imaging of a wide range of astronomical objects from planets to quasars (Figure 7.13). The acquisition and guiding systems available at many telescopes usually employ either SIT or SEC tubes, often with extra intensification, operating in the imaging mode. The principal drawback in using electron beam readout tubes for imaging is the beam bending encountered in the vicinity of high contrast regions of the image. The large change in positive potential over a small distance on the rear of the target means that there is a transverse electric field in the plane of the target to deflect the negatively charged read beam away from its correct landing point. It is hoped that the solid state imaging devices described in Chapter 8 of this book will eliminate this problem.

7.8 Image-dissector scanner

A different approach to building an imaging detector has been taken by Lick Observatory in developing the image-dissector scanner, initially used exclusively for multichannel spectroscopy but now also capable of two-dimensional imaging. The principal functional modules of the instrument are shown in Figure 7.14.

Three cascaded electrostatically focused image tubes provide an optical gain of 10^5 between the first photocathode and the output phosphor of the third image tube. The output phosphor is fibre optically coupled to another tube called the image-dissector tube in which electrons from the photocathode are directed by combined electrostatic and magnetic fields towards a metal plate in which is a small rectangular aperture leading to an electron multiplier. Only those electrons originating from one small area of the photocathode will have the correct trajectory to pass through the aperture. All other electrons will hit the plate and be returned to

Figure 7.14. The image-dissector camera tube coupled to a three-stage intensifier.

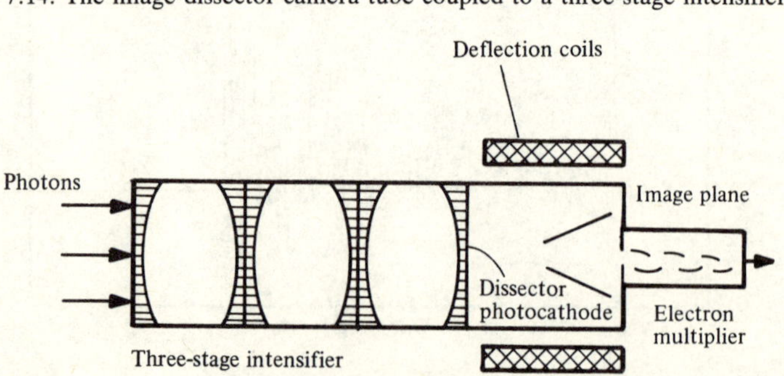

ground. The magnetic field is produced by a set of focus and deflection coils in a similar way to the more usual television-type camera tubes. The 'correct' electron trajectory and hence area of the photocathode seen by the electron multiplier is scanned in a normal raster pattern with the size of the exit aperture defining the spatial resolution. At Lick, an aperture of 37×250 μm is used in the spectroscopy mode in which two parallel spectra representing sky and star plus sky are scanned over a distance of 2 cm each at 2048 channels per spectrum. The pixel size is therefore 10 μm and spatial resolution 50 μm.

The image scanner tube does not possess storage akin to a camera target so that the whole of the photocathode must be scanned within the typical phosphor decay time of 5 milliseconds. This is equivalent to one million pixels per second and is rather a high data rate for faint object detection. However, the total gain from first photocathode to last electron multiplier dynode is so high that primary photoelectrons can be detected above the readout noise. Objects as faint as $V = 21$ magnitudes have been observed with the Lick 3.05 m telescope. The system is not truly photon counting as, depending on the image tube gain, up to ten output pulses are produced for each primary photoelectron. Nevertheless, the image-dissector scanner is another heavily used and very successful low light level detector.

References for Chapter 7

Biberman, L.M. & Nudelman, S. (eds.), 1971. *Photoelectronic Imaging Devices*, Plenum, New York.

Glaspey, J.W. & Walker, G.A.H. (eds.), 1973. *Astronomical Observations with Television-Type Sensors*, University of British Columbia.

McMullan, D. & Morgan, B.L. (eds.), 1978. *Proceedings of Seventh Symposium on Photoelectronic Image Devices*, Imperial College, London.

8

Solid state imaging detectors

Since the mid 1960s there has been an attractive alternative to the electron beam imaging detectors just described, namely the charge coupled device (CCD) and its near relative the charge injection device (CID). These solid state devices offer the obvious advantages of greatly reduced size, weight and power consumption together with the astronomically more important advantages of high quantum efficiency and low readout noise. There are as always penalties to be paid, the principal one being the large number of flaws in the recorded image. This has prevented these devices being used extensively in commercial television cameras, which has in turn kept the price of large format devices rather high. However, these devices are now in use at many observatories and are rapidly becoming the most important class of imaging detectors.

The CCD is by far the most important of these devices and will be described in considerable detail before a brief look is taken at the other types of solid state imaging detectors.

8.1 Charge storage in MOS capacitors

The principal component of a CCD is a charge storage capacitor which is very similar to that found in a metal oxide semiconductor (MOS) transistor. The basic construction of the MOS capacitor is shown in Figure 8.1. A silicon dioxide (SiO_2) insulating layer is grown onto a p-type doped silicon substrate. The insulating layer is etched away to a depth of about 0.1 μm and a metal electrode called a gate is deposited. If the gate is biased positively with respect to the substrate, the majority carriers, or holes, are repelled from the $Si-SiO_2$ junction and a depletion layer forms. As the gate voltage, V_G, is increased the depletion layer moves deeper into the substrate. Once V_G exceeds a couple of volts, the $Si-SiO_2$ junction becomes sufficiently positive with respect to the bulk substrate that any free electrons present are attracted to the junction and form an inversion layer. In an MOS transistor, this inversion layer is the channel through which electrons pass from the source to the drain. In an uncooled

Figure 8.1. (a) Basic construction of a metal oxide semiconductor capacitor. (b) Formation of a depletion layer under reverse biasing conditions.

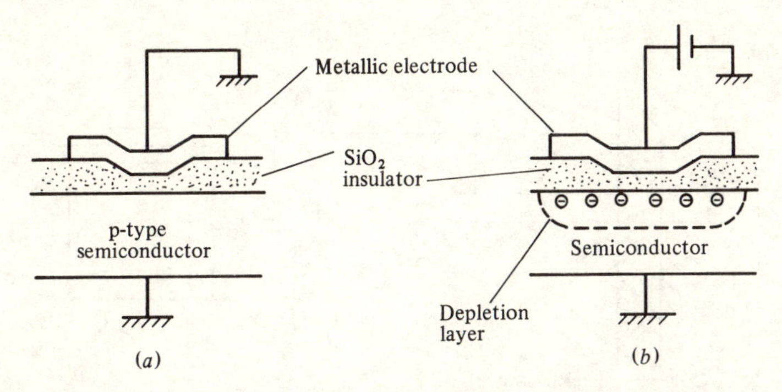

MOS capacitor, thermally produced electrons will rapidly flow into the junction. However in a cooled device this inversion layer will not form, unless free electrons are generated by photon bombardment.

The depletion layer can also be thought of as a reservoir or potential well into which charge can be deposited and later retrieved. Figure 8.2 shows the energy level diagrams for empty and partially filled potential wells. In the absence of free electrons, most of the gate potential is distributed throughout the substrate and the potential well has a depth of approximately V_G. As the $Si-SiO_2$ interface, and potential well, is filled with electrons, the interface potential drops and the energy bands level out. Eventually the Fermi level in the substrate equals the potential of the bottom of the conduction band whereupon the well is full and almost the whole potential difference V_G is across the SiO_2 insulator. It must be stressed that although the potential well concept is very useful and will be used throughout the rest of this chapter, the electrons are in reality stored from the $Si-SiO_2$ interface downwards into the substrate and not from the bottom of a well upwards. Of course this model also applies to n-type semiconductors with V_G being negative and the stored minority carriers being holes. However in this chapter, the minority carriers are assumed to be electrons. The gate electrode is not always a true metal but quite often made of polysilicon which is a form of silicon having metallic properties. This enables the whole device to be silicon thereby reducing contamination problems and improving optical transmission into the substrate.

8.2 The surface channel CCD

A charge coupled device consists of an array of storage and image cells connected so that charge can be transferred from one cell to the adjacent

Figure 8.2. Energy level diagram and potential well representation of a MOS capacitor: (a) with V_G applied and well empty; (b) with some charge at the oxide–semiconductor interface.

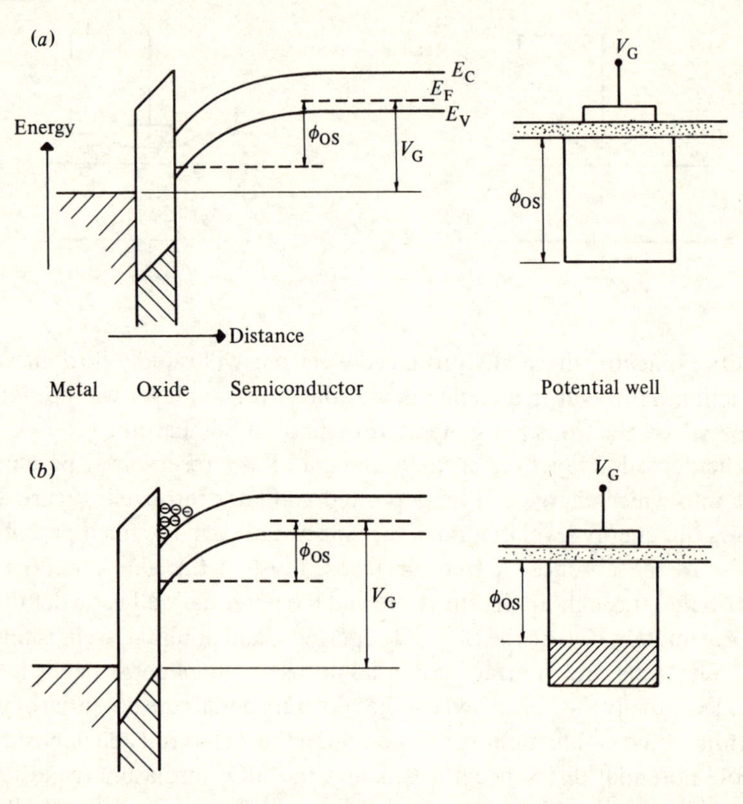

cell and finally to an output amplifier. For the surface channel CCD, the cells consist of the simple MOS capacitors just described in which charge is stored at the surface of the p-type semiconductor. The easiest CCD configuration to understand is the three-phase CCD shown in Figure 8.3. At time t_1 the gate electrode $\varnothing_1$ is positive and the other gates $\varnothing_2$, $\varnothing_3$ are at ground potential. The substrate is biased slightly negative with respect to ground to ensure that the $Si-SiO_2$ interface is free of holes. This greatly improves the efficiency of the charge transfer process described next. The potential well under $\varnothing_1$ is partially filled with electrons which could have been either injected into the device through an input gate or created in the bulk silicon through absorption of incident photons. At time t_3, gate $\varnothing_2$ also goes positive creating a well which extends under both electrodes. The stored charge redistributes itself uniformly underneath both gates in a similar way to that of water seeking its own level. At

Figure 8.3. Basic operation of a three-phase CCD. (a) The potential well charge distribution at times t_1, t_2 and t_3. (b) The three-phase voltage clocking waveform.

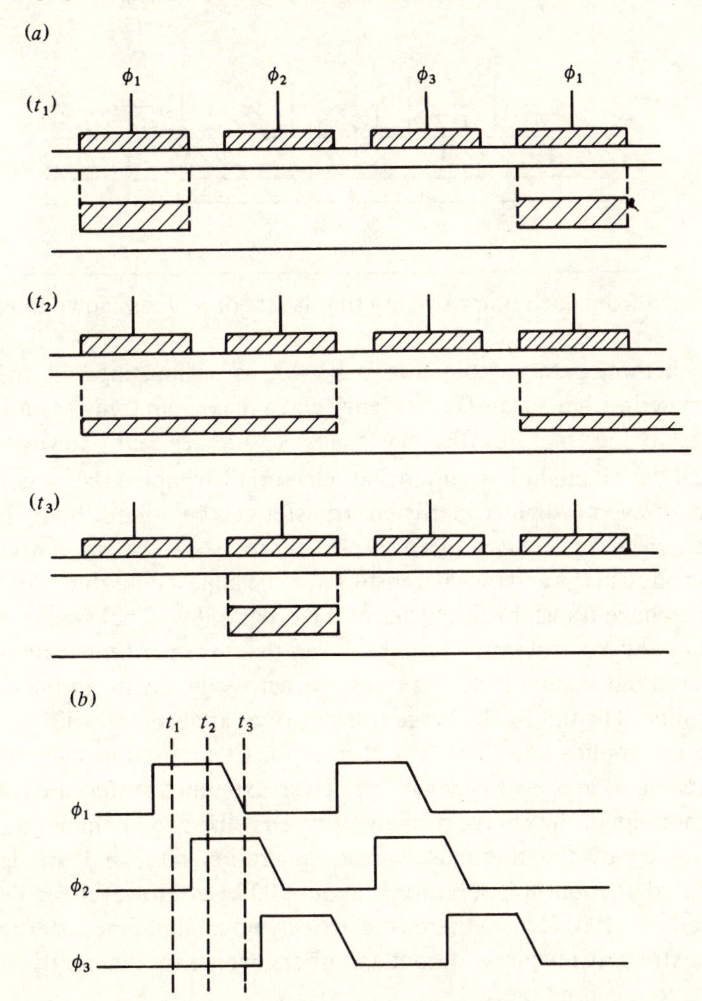

time t_3 the voltage on gate $\emptyset_1$ is ramped down to zero leaving all the charge in a potential well beneath gate $\emptyset_2$. The gate voltage has to be ramped down rather than cut off suddenly to ensure that no charge splashes out of the well and suffers recombination. A similar series of three steps will then move the charge from $\emptyset_2$ to $\emptyset_3$ and so on until the charge reaches the output gate. Note that the three-phase system is bidirectional, a feature of importance in certain applications. To ensure efficient charge transfer between gates, the interelectrode spacing must be small. This can be achieved by employing overlapped polysilicon gates

Figure 8.4. Overlapping gate structure used in a three-phase surface channel CCD.

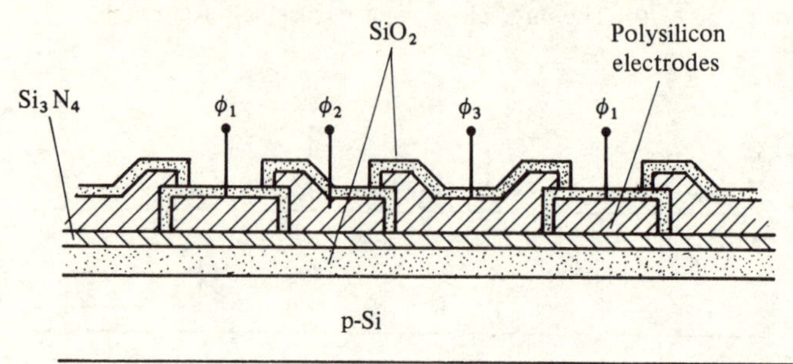

insulated from each other by very thin layers of SiO_2 as shown in Figure 8.4.

Each three-gate combination is known as an imaging cell and has associated with it a transfer efficiency and a maximum transfer rate. In a real CCD the transfer efficiency is close to 100 per cent. However as a typical device might have up to 500 cells serially coupled the cumulative effect of even a slightly inefficient transfer can be quite serious. In the extremely inefficient case of say 99 per cent transfers, the image would be so smeared at the end of 500 transfers as to be almost unusable. The latest devices have transfer efficiencies of the order of 99.99 per cent or better so that any particular pixel will have less than one per cent of its charge smeared out behind it during transport across the device to the output amplifier. The inter-cell charge transfer rates are generally different for different applications. The typical rates for CCDs used in studio television cameras are several megahertz. The maximum transfer rate is set by the mobility of the charge packets between cells. The minimum transfer rate is set by the thermal electron generation rate, and for devices operated at room temperature is about 100 kHz. However for devices cooled to −100 °C or so there is effectively no minimum transfer rate so that extremely low noise output amplifiers, which are inherently slow in operation, can be used.

So far only the three-phase CCD has been discussed. The principal practical drawback is the requirement for three overlapping clock drive pulse trains. A somewhat simpler arrangement is the two-phase device shown in Figure 8.5. One set of electrodes, made of polysilicon, is embedded in the SiO_2 insulator such that the potential wells under the electrodes are deeper than under the gates on the surface of the insulator. As a result a two-tiered asymmetric well is created in the bulk p-type silicon. In the figure the electrodes connected to phase $\varnothing_2$ are more positive than those connected to phase $\varnothing_1$ and the deeper well contains

Figure 8.5. Operation of a two-phase CCD. (a) Physical electrode structure. (b) Movement of charge. (c) Voltage clocking waveform.

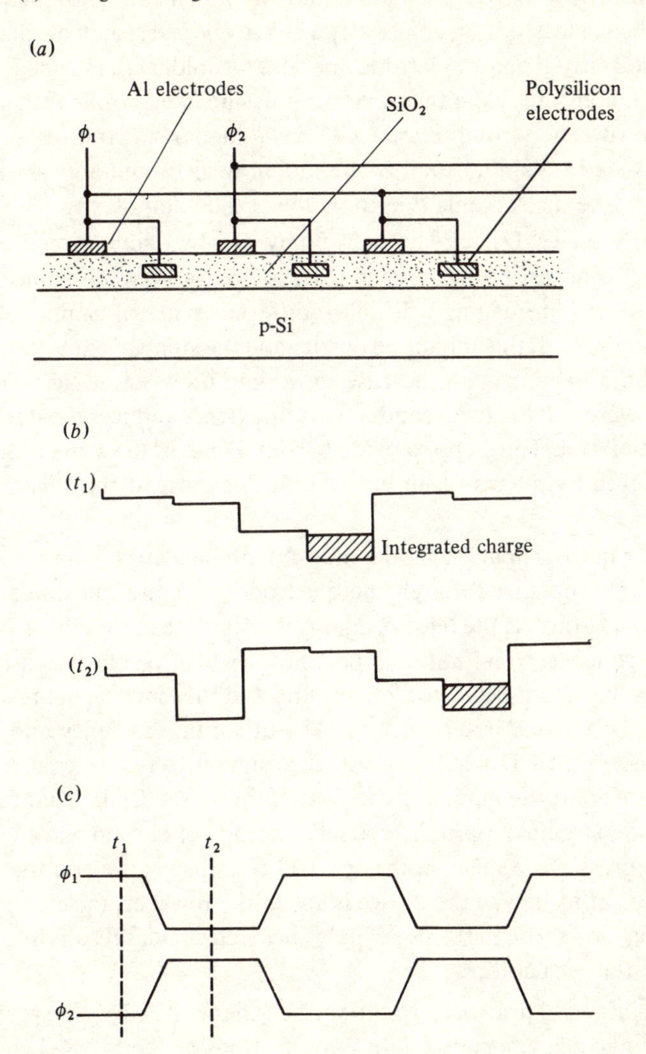

some integrated charge. At a later time, t_2 the relative potentials of $\emptyset_1$ and $\emptyset_2$ are interchanged and the charge moves to the deeper well now under the next $\emptyset_2$ electrode to the right. The price to be paid for the decreased complexity of clocking is a reduction in cell storage capacity compared to the three-phase electrode structure. The two-phase structure can be operated with a single clock that is driven above and below a constant potential applied to the other phase.

A recent development has been the elimination of the two layer metallisation in favour of ion implantation in which the constant potential polysilicon electrode is replaced by a heavily doped region of silicon. This so-called virtual phase electrode operates by holding the surface potential under it constant while the adjacent polysilicon electrode is clocking up and down. The virtual phase CCD has a higher spectral response, uniformity and reliability than conventional devices and appears ideal for space borne instruments due to its lower gate complexity.

Nearly all CCDs used in astronomy are two-dimensional arrays in which it is necessary to prevent charge leakage between adjacent channels or rows of imaging cells. Two different types of channel stops are commonly used, the implanted barrier and the stepped oxide barrier. The implanted barrier is an increase in doping between adjacent channels which increases the local conductivity and hence reduces the depth of the potential well. The stepped oxide barrier achieves the same reduction in well depth by means of an increase in thickness of the silicon dioxide layer.

In the normal imaging mode, photons are incident on the front face of the device, passing through the electrode structure and into the bulk silicon substrate. If the photon energy is greater than the silicon band gap energy, one electron–hole pair per photon will be produced with the hole passing through the substrate to ground and the electron being attracted to the the nearest potential well. The quantum efficiency and spectral response of a CCD would therefore be similar to that of a silicon diode were it not for the gate electrode structure on the CCD front face. The electrode structure partially reflects photons of certain wavelengths so that depressions appear in the spectral response curve and the average quantum efficiency of the device is lowered. However, the elimination of metallisation through the use of polysilicon electrodes results in an almost transparent structure.

The principal drawback to the surface channel CCD is the presence of trapping states associated with imperfections at the Si–SiO_2 interface. Remembering that electrons are stored from the interface downwards it can be seen that the trapping and hence loss of transfer efficiency will be highest for the smaller charge packets which correspond to the fainter astronomical images. To alleviate the problem, the CCD can be permanently illuminated with a steady low level light source so that the surface states are always filled. This artificial bias or 'fat zero' as it is generally called greatly improves the linearity at low light levels and can be removed during subsequent signal processing.

8.3 Buried channel CCDs

The presence of surface states and the consequent entrapment of charge in the surface channel CCD significantly reduces its efficiency at low light levels. The solution is to store and move the charge packets in the bulk semiconductor away from the $Si-SiO_2$ interface. This has been accomplished by the development of the buried channel CCD, illustrated in

Figure 8.6. Buried channel three-phase CCD. (a) Basic construction showing the extra n-doped layer introduced between the SiO_2 insulating layer and the p-doped substrate. (b) Potential distribution through the silicon showing the formation of an electron well at the n-type p-type junction.

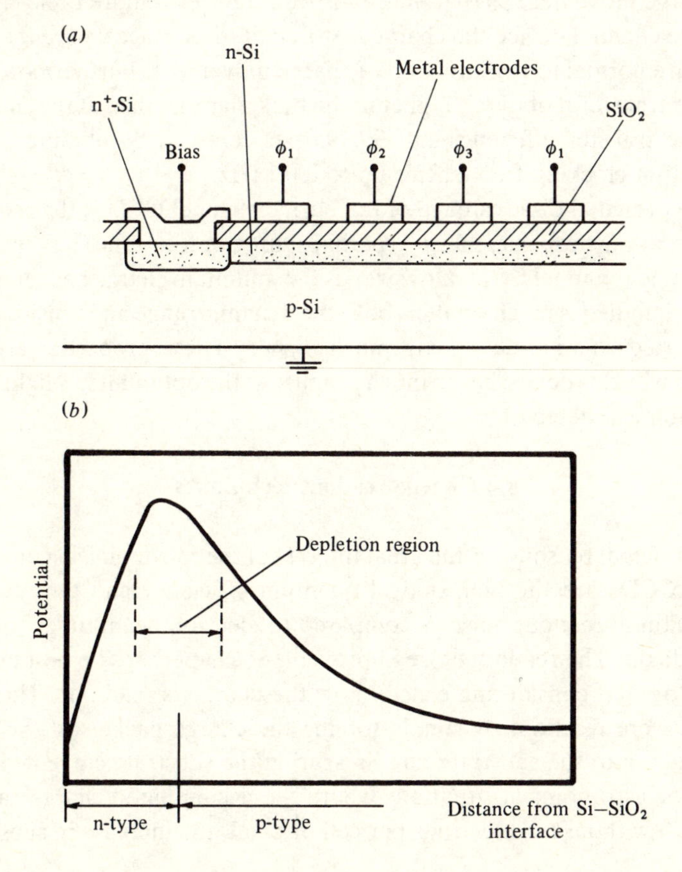

Figure 8.6. The substrate immediately under the silicon dioxide is n-doped by means of ion implantation. If a voltage more positive than any of the gate voltages is applied to this n-type layer, a depletion layer will form at the p-n junction where the substrate potential is a maximum. This is also shown in the figure. The exact value of the potential and hence the depth of the charge storage well depends on the electrode gate potential immediately above. The wells will, as before, be deepest under the most positive gates so that spatially separated storage wells are still formed, but they are now far from the oxide–substrate surface. The movement of charge throughout the device is similar to that for surface channel devices but is more efficient for two reasons. Firstly the number of trapping centres in the substrate bulk is much less than at the surface so that the transfer at low charge levels is much improved, and secondly the electric field lines between adjacent electrodes are nearly parallel to the direction of charge movement so that the cell-to-cell transfer is more rapid. In the surface channel device the charge is stored at the surface where the field lines are normal to the direction of charge movement. Furthermore, the charge transfer mobility is higher in the bulk material than at the surface. Charge transfer efficiencies of 99.995 per cent can be obtained, with maximum clocking frequencies up to 100 MHz.

The principal disadvantage of the buried channel CCD is the reduced charge storage capacity which is perhaps three or four times less than for the surface channel CCD. However as the minimum charge packet which can be handled is much smaller, both the dynamic range and sensitivity of the buried channel device are much higher. These properties are the reason why this device has so much promise as the optimum low light level astronomical detector.

8.4 Charge readout techniques

The principal reasons for the great interest of the astronomical community in CCDs are the high optical quantum efficiency and the ten-fold reduction in readout noise as compared to electron beam tubes such as the vidicon. The readout noise improvement can perhaps be best understood by first considering exactly how the charge is read out. The first CCDs were read out by simply forcing the charge packets at the final electrode into the substrate and measuring the substrate current. However for astronomical situations where the accumulated charge can be only a few thousand electrons per cell or pixel, the increase in substrate current would be very small indeed. A much better approach is to convert the charge packet to a voltage increment according to

$$\Delta V = \Delta Q/C$$

where ΔV is the voltage increment, ΔQ is the size of charge packet and C is the effective capacitance to ground of the sensing electrode. The interelectrode capacitance to ground of a vidicon-type tube can be several tens of picofarads with the result that ΔV is very small. However in the case of CCDs it is possible to arrange the capacitance to be much less than a picofarad and moreover the first stage of the video amplifier can be built immediately adjacent to the sensing electrode, further lowering the read-out noise. There are two widely used charge sensing methods in current use.

The first method is known as floating diffusion sensing and is illustrated in Figure 8.7 for the case of a three-phase surface channel CCD. Electrode $\varnothing_3$ is the usual last electrode of a three-phase cell and is followed by an extra electrode, the output gate, and a reverse biased diode which is manufactured by diffusing an n-type region into the p-type substrate. This output diode or diffusion is connected to the input of an on-chip MOS transistor amplifier and also to a reset MOS transistor. Charge is clocked from $\varnothing_3$ to the output gate, $\varnothing_{OG}$, and then into the output diode. This reverse biased diode is equivalent to a small capacitor and thus will experience a change in potential on arrival of the charge packet. The voltage change is amplified by the MOS transistor and passed on to the

Figure 8.7. Method of charge sensing known as floating diffusion sensing. The capacitor, C, represents the parasitic capacitance of the n-type output diffusion. Charge is passed to this diffusion by control gate $\varnothing_{OG}$.

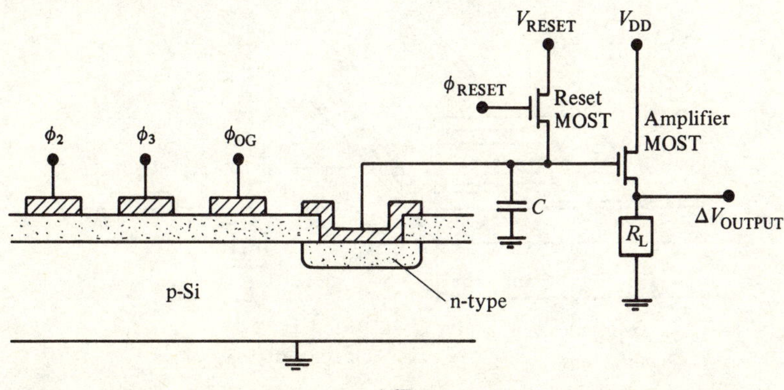

following video processing chain. A reset pulse, $\varnothing_{\mathrm{RESET}}$, then turns on the reset transistor and resets the output diode to the potential, V_{RESET}, which is the potential against which voltage changes are measured. The major source of non-linearity with this method is the change in diode capacitance with diode potential. This could cause considerable problems with high contrast images. However, it can be minimised by loading the diode capacitance with a larger external capacitance, C, which can be most easily implemented by designing the MOS amplifier to have a large capacitance to ground.

For the low signal levels found in faint object astronomy, the feed-through of the clocking and reset pulses into the output transistor can be larger than the optically generated signal. A special signal processing technique known as correlated double sampling has been developed which essentially eliminates the effect of this pulse feedthrough. This technique relies on sequentially switching two capacitors in the signal chain in such a way that only the difference in signal due to the optically generated charge is detected. More details can be found in the references at the end of this chapter.

The sensing method just described is destructive in the sense that the charge packets are destroyed in the readout process. Another common sensing technique which is non-destructive is floating gate sensing. The advantage of a non-destructive readout is that multiple reads can be employed to reduce the readout noise. The basic electrode arrangement is shown in Figure 8.8, again for a three-phase surface channel CCD. Interspersed between two normal CCD electrodes is an electrode which is

Figure 8.8. Floating gate charge sensing method.

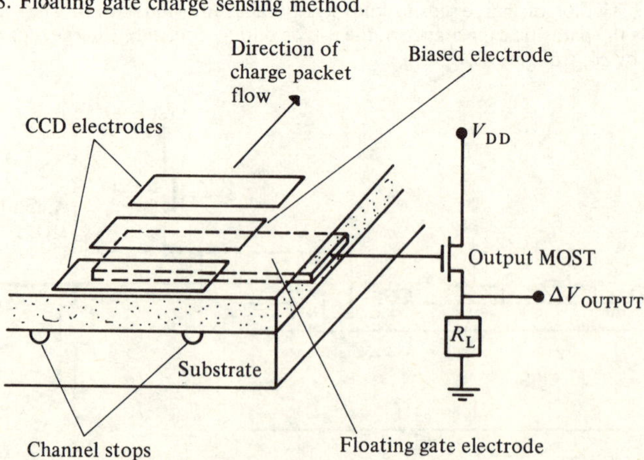

held at a fixed bias. Deposited in the silicon dioxide layer under this biased electrode is an extension of the gate of an on-chip MOS transistor. Charge packets are transferred along the device in the normal way. When the charge packet passes under the biased electrode it produces a change in the floating gate potential through capacitive coupling. The equivalent gate capacitance is smaller than the floating diffusion capacitance so that the charge sensitivity is higher. Another advantage is the elimination of reset pulses and their associated noise.

The non-destructive nature of the readout enables multiple readouts to be used to increase the signal-to-noise ratio as illustrated in the three-phase distributed floating gate amplifier (DFGA) shown in Figure 8.9. The charge packets are clocked into the first floating gate amplifier as described previously. The output of this amplifier is passed to a separate output register. The charge packets are then clocked into the next floating gate amplifier. The output of this amplifier is also passed to the output register. However, since the output register has been clocked at the same time as the main CCD chain, the output register stage connected to this second floating gate amplifier already contains an amplified copy of the charge packet, and thus the signal in the output register is doubled. At the next floating gate amplifier another amplified copy of the charge packet is added to the output register. By the time the end of the output register is reached each charge packet will have been read as many times as there are stages in this register. The accumulated charges are collected at the end of the output register and further amplified before leaving the chip. If there are N stages the output register, the signal-to-noise ratio will be improved

Figure 8.9. Distributed floating gate amplifier.

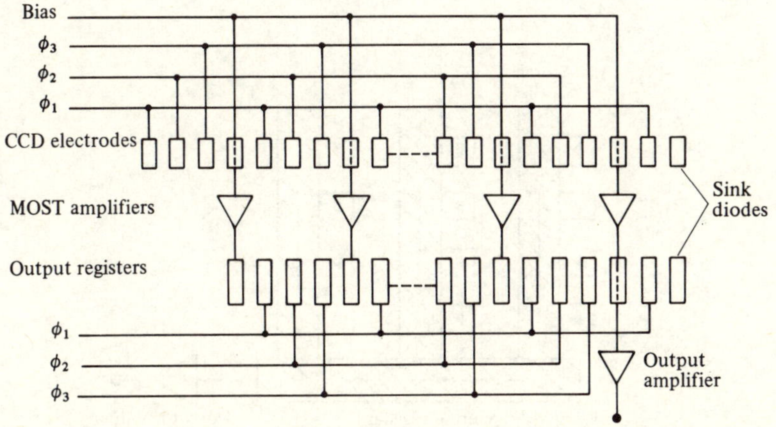

by $N^{1/2}$. In this way readout noise levels as low as 20 photoelectrons per pixel have been obtained.

8.5 CCD architectures

So far the various constituent parts of surface and buried channel CCDs have been considered. This section will discuss the different cell configurations possible for use as a two-dimensional imaging detector and the influence a particular configuration has on the sensitivity, spectral response and resolution of the CCD.

There are two principal arrangements of cells in common use, frame transfer and interline transfer, both shown in Figure 8.10. In the frame transfer architecture, the cells are arranged in an array of vertical columns and horizontal lines. One half of the array is the imaging section in which

Figure 8.10. Charge coupled device arrays employing (a) frame transfer architecture, and (b) interline transfer architecture.

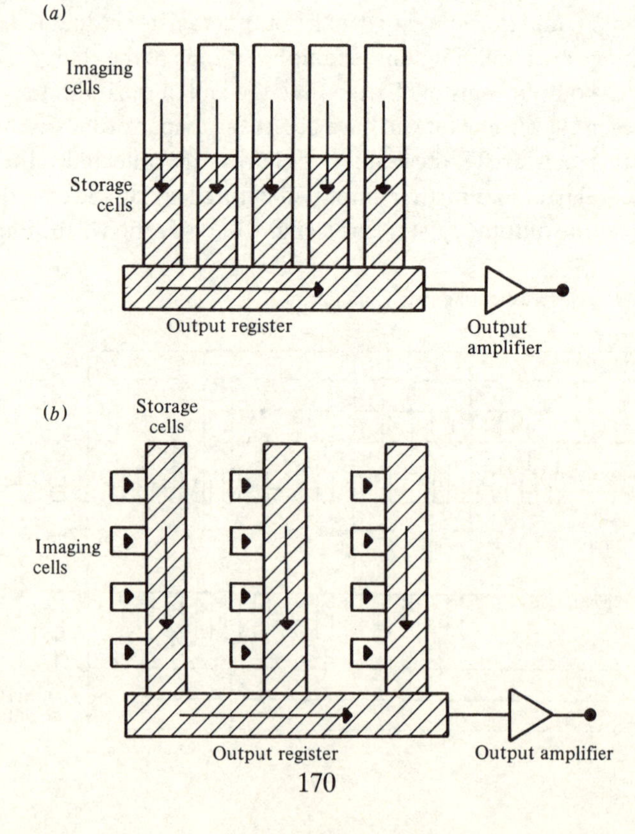

charge packets are built up as described previously and the other half is a storage section. After the required integration time, the integrated charge packets are moved to the storage cells which are shielded from optical illumination by masking or a metal overlay. This storage array is then moved line by line into the output register and passed to the output amplifier. While one image is being read out from the storage section another can be integrated in the image section. During the charge transfer the imaging section remains exposed to light. The movement of charge between the two sections must be fairly rapid otherwise the continuous exposure of the imaging cells will cause a slight smearing of the image as charge is moved through them. The movement along the output register can be slower to minimise readout noise.

In the interline transfer architecture the columns of imaging cells are separated by optically shielded line transfer registers as shown in Figure 8.10b. The charge distribution is built up as usual in the imaging cells for the required integration time and then all cells are simultaneously shifted rapidly sideways into the transfer registers. This sideways shift takes place in only a few microseconds so that image smear is virtually non-existent. The charge packets are then clocked down the transfer registers into the output register and through to the output amplifier. The principal disadvantage with this architecture is that the imaging section is not continuous but contains equal numbers of sensitive imaging and insensitive transfer cells. The effective quantum efficiency of the basic silicon substrate is therefore reduced, by the ratio of imaging area to transfer cell area. The maximum quantum efficiency of the interline transfer CCD is only about 30 per cent as opposed to the 60 per cent figure typical for frame transfer devices. The quantum efficiency and spectral response are further modified by the electrode structure through which the photons must pass before being absorbed in the silicon substrate. Since the various parts of the electrode structure have dimensions comparable with the wavelengths of visible light the quantum efficiency of the device will be reduced at certain wavelengths. The spectral response curve is no longer a smooth slowly varying curve but takes on a rippled appearance.

The regular spacing of the cells making up the imaging area of the CCD gives rise to a maximum spatial frequency beyond which image detail is ambiguous. This ambiguity, or aliasing as it is usually called, arises because not enough data samples are taken to define the image structure uniquely. Aliasing is a common problem with all imaging devices constructed from discrete cells. For frame transfer arrays the imaging cells are usually equal sided so that there is just one modulation transfer function for the array. However, for the interline transfer array, interspersing imaging cells with transfer cells produces two orthogonal modulation functions and in general a worse aliasing problem.

Provision must be made for preventing charge which exceeds the maximum cell capacity from spilling over and spreading sideways through the CCD, otherwise optical overload from either too bright an object or too long an integration could contaminate adjacent cells. This effect is called blooming, and one cure is to introduce a slight negative bias to the adjacent electrodes in the cell so that the excess negative charge will recombine with the positive charges formed under the negative electrodes. Another approach is to diffuse or implant anti-blooming drains in between the imaging cells such that any excess charge which overcomes a potential barrier is conducted away safely to ground.

These different architectures may be illustrated by reference to two commercial CCDs that have been successfully used as astronomical detectors, the CCD211 manufactured by Fairchild Corporation and the MA357 by GEC. The CCD211 is a buried channel two-phase device organised in an array of 244 horizontal lines and 190 vertical lines using the interline transfer architecture. The image sensing elements are 14 μm $\times$ 18 μm with the adjacent transport register cells being 16 μm $\times$ 8 μm. Consequently the effective quantum efficiency is only about 20 per cent at best. The spectral response peaks at 850 nm with useful response from approximately 450 to 1150 nm. There is a small amount of spectral rippling due to the electrode structure. The standard output is a floating gate amplifier which gives a system noise level of about 100 electrons rms per pixel per readout. However a distributed floating gate amplifier is available and reduces the noise to about 20 electrons rms. The saturation charge per cell is 250 000 giving a dynamic range of 10 000 to 1 over which the device is perfectly linear.

The MA357 is a three-phase buried channel device organised as a 385 $\times$ 576 pixel frame transfer array. The electrodes are polysilicon and the device is usually used front illuminated. The pixels are each 22 μm square and are fabricated with no dead space between adjacent pixels. As a result the quantum efficiency is good showing a peak quantum efficiency of 40 per cent at 800 nm as shown in Figure 8.11. The readout noise is typically 20 electrons rms. The full well capacity of each pixel is about 500 000 electrons giving a dynamic range of 25 000 to 1. The dark current has been measured at 0.01 events per second per pixel at 173 K and was undetectable at 105 K. As with all devices, there is a slight loss of red

Figure 8.11. Quantum efficiency of the GEC device MA357, working at a temperature of 105 K.

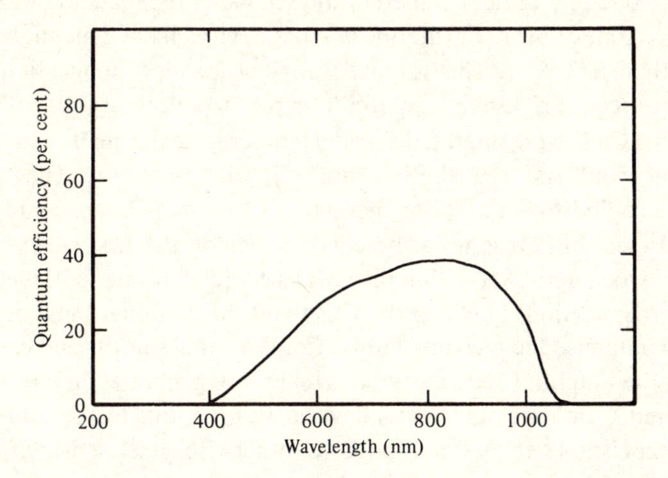

sensitivity but Figure 8.11 shows that even when cooled, useful response can be obtained up to 1.05 μm.

As with other classes of detectors the CCD can be intensified. Photons incident on an intensifier tube photocathode release electrons which are accelerated towards the CCD. Rear electron bombardment of the CCD is to be preferred in the interests of minimising damage to the electrode structure. Intensification is not, however, often used as the lower quantum efficiency and reduced spectral range of the photocathode are definite disadvantages. In addition, the readout noise of a CCD fitted with a distributed floating gate amplifier is so low that the extra gain of the intensifier does not significantly improve the signal-to-noise ratio.

8.6 Practical CCD operation

Charge coupled devices are finding increasing astronomical use as efficient detectors for both spectroscopy and imaging. Medium resolution spectroscopy of quasars and faint galaxies to determine their redshifts requires a detector with high pixel positional stability and very low internal sources of noise so that the required wavelength accuracy can be reached in an integration period much less than one night. The same applies to high resolution spectroscopy of brighter objects. Imaging of faint galaxies requires a detector which is linear over a wide dynamic range in addition to being stable and of low noise. Current CCDs have two great advantages over television-type detectors such as the silicon vidicon. Firstly, the pixel position and size are stable and independent of the electrode operating parameters and secondly, the readout noise is at least five times lower. Furthermore most devices have dynamic ranges larger than 1000 to 1. This is quite sufficient for most applications.

Faint object work usually requires integration periods of one to several hours. If a CCD were operated at room temperature thermally generated electrons would saturate the imaging cells within seconds. The device must be cooled to −100 °C or thereabouts to eliminate the buildup of thermal electrons. It is usual practice to include the first stage of the external video amplifier within the CCD's cooled housing, both because the interconnections between the CCD and the amplifier must be kept short to minimise the pick up of noise from external sources and because cooling the amplifier reduces the thermal noise generated in its various components. In some of the early devices the operating temperature had to be kept stable to better than 1 °C otherwise pixel sensitivity was somewhat variable. However with changes in manufacturing procedures the exact temperature is no longer critical.

A typical CCD frame is taken by first exposing the device to a uniform bright light which saturates all pixels. The device is then read out several times to empty all pixels to the same very low level. The image is then allowed to fall on to the device for the required integration period. The CCD array is then read out slowly line-by-line in a typical time of 10 seconds. The individual pixel analogue signals are converted to an equivalent digital count and stored in some kind of digital memory where they can be readily accessed by a suitable computer. Since a typical CCD array has about 10^5 pixels, rather large computers are required for data manipulation. Virtually every CCD available shows both smooth variations in sensitivity across the device and discrete pixel defects. A lot of effort has gone into removing these imperfections. The easiest data correction to apply is that required by the variation in sensitivity across the device, which amounts to typically 10 to 15 per cent. This flat-fielding is achieved by taking a CCD exposure of a diffuse uniform source such as the defocused moon and then dividing the data frame by it, pixel by pixel.

Following this step the night sky brightness contribution is removed by subtracting a frame taken of the night sky under the same conditions as that of the data frame. This frame subtraction is very effective for CCDs because of the excellent device linearity and pixel stability.

Much more difficult to remove are the various discrete manufacturing defects. It is sometimes found that no matter how low the device is cooled, certain pixels appear to be completely saturated with thermally generated electrons. These so called 'hot spots' are usually eliminated in the data processing by replacing the saturated pixel with an intensity interpolated from adjacent pixels. A more serious problem is the presence of 'dead' pixels which have greatly reduced detection and transfer efficiencies. The bad pixel is often in such a position in the array that the movement of good charge packets through the bad cell seriously contaminates the data, showing up later on the reconstructed image as a spurious line. It is not usually possible to remove this defect completely. Bright stellar images are often seen to leave a 'star trail' behind them giving them the appearance of comets. This is caused by cell transfer inefficiency. For example, if the sky background were equivalent to 100 photons per pixel and a star had been overexposed giving one million photons per pixel, a cell transfer efficiency of 99.99 per cent would double the effective sky background of every pixel through which the overexposed stellar image were passed.

Cooled devices often still show random events similar to thermal noise spikes. These are thought to be caused by secondary particles from cosmic rays. A typical device shows a cosmic ray event rate of about one per minute over the whole array and represents the most serious problem in long term integrations since the cosmic ray events are difficult to distinguish from stellar images. One observational approach is to divide a required integration time into several smaller time segments and to ignore any apparent stellar image which does not appear on all segments. The event rate can be reduced by an order of magnitude by drastically thinning the device to reduce the volume of silicon in which cosmic rays can be trapped. However the mechanical rigidity is seriously weakened and special manufacturing care must be taken to keep the device flat. Thinning also enables back illumination which eliminates the spatial interference of the electrode structure. It has the further effect of improving the blue response at the cost of losing some of the near infrared response. An alternative approach to thinning is to build the CCD on an epitaxial layer of low resistivity silicon within which cosmic ray generated charge pairs quickly recombine before reaching the CCD electrodes. This technique gives a device which has low cosmic ray sensitivity but high mechanical rigidity. The GEC MA357 is such a device and has a measured cosmic ray event rate of some two to three per minute over the entire CCD. However this approach does not give the enhanced blue response of true thinning.

Figure 8.12 shows two images of the Seyfert galaxy, ESO G144-195, taken with the Cambridge MA357 CCD system on the 4 m telescope at

Figure 8.12. Photographs of the Seyfert galaxy, ESO G144-195, taken with the Cambridge CCD system. (a) Raw data of 5 minute *R*-band exposure. (b) Data corrected for detector sensitivity non-uniformities.

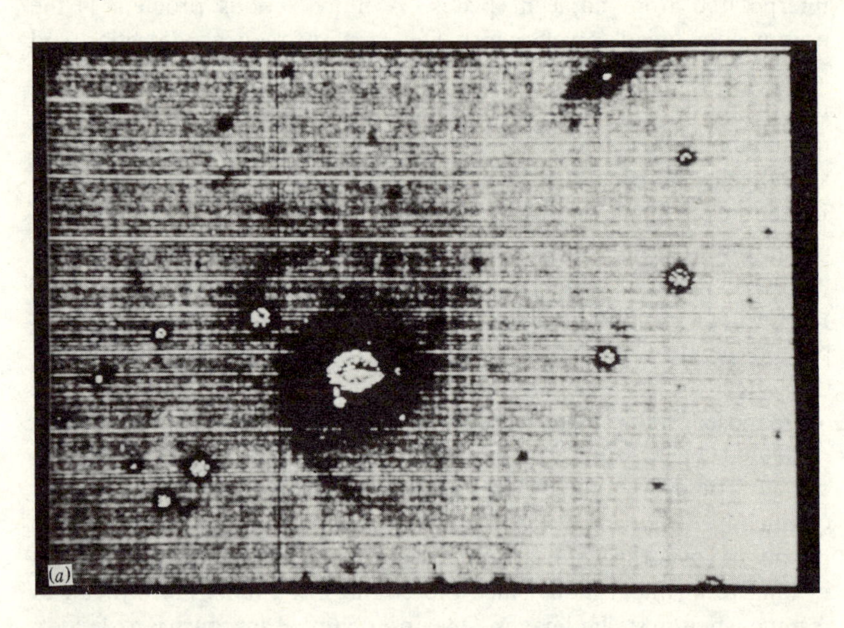

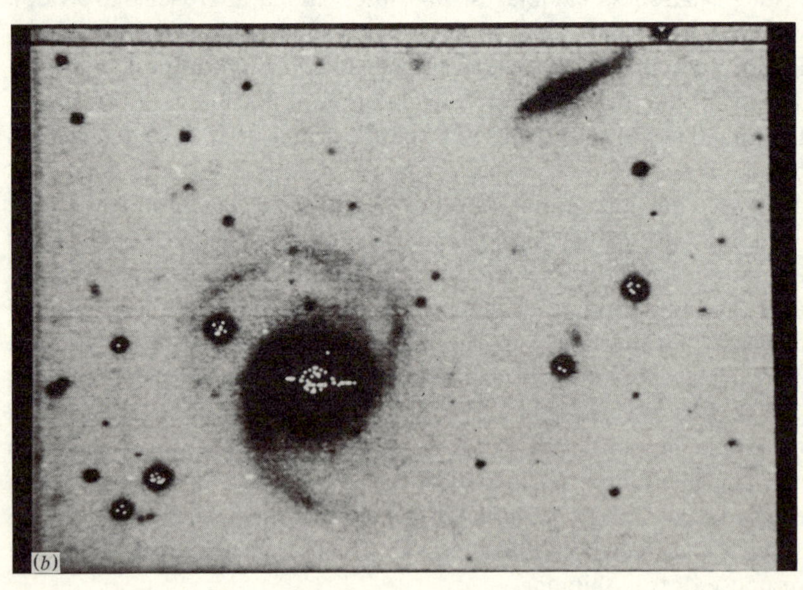

the Anglo-Australian Observatory. The top image is the 'raw' output from the CCD and shows several minor defects. The lower image is the 'cleaned-up' image which has been processed by the techniques described above.

The Cambridge group has promoted a technique known as drift scanning which largely eliminates the problems associated with flat fielding that occur when operating close to the sky background limit. With this technique the CCD is clocked out at a rate of a few rows per second as soon as the integration period starts. The image smearing that would normally result is avoided by mechanically moving the optical image in synchronism with the already accumulated charge distribution. The signal then builds up linearly until one full frame has been read out. All subsequent rows read consist of image pixels that have dwelt for equal times on every device pixel on a single column of the CCD. Therefore all image pixels read out from a particular column have been detected with the same mean device pixel efficiency and should show no response non-uniformity along the columns of the output image. Any remaining column-to-column non-uniformities can be removed by either measuring sky levels between objects or requiring that the output image should have no structure parallel or perpendicular to the device columns. This technique gives about one magnitude improvement in sensitivity as compared to conventional exposure and flat fielding procedures. The main disadvantage of drift scanning is the time required for the initial buildup of charge which represents one full frame integration time. This disadvantage is a maximum when the area of sky being observed is equal to the area of the CCD.

8.7 Charge injection device

Several alternative solid state imaging devices have been developed over the past decade and while none has proved more popular than the CCD, each one does have unique characteristics which can be important for certain astronomical observations. The device closest to the CCD in mode of operation is the charge injection device or CID. The basic mode of operation can be understood by reference to Figure 8.13 which shows the essential details of a 4 × 4 element CID array. Each cell consists of a pair of adjacent MOS capacitors similar to those of the surface channel CCD (Figure 8.1). One of the two electrodes is held at high voltage so that a potential well is formed beneath it capable of storing optically generated

Figure 8.13. Organisation and operation of a charge injection device.

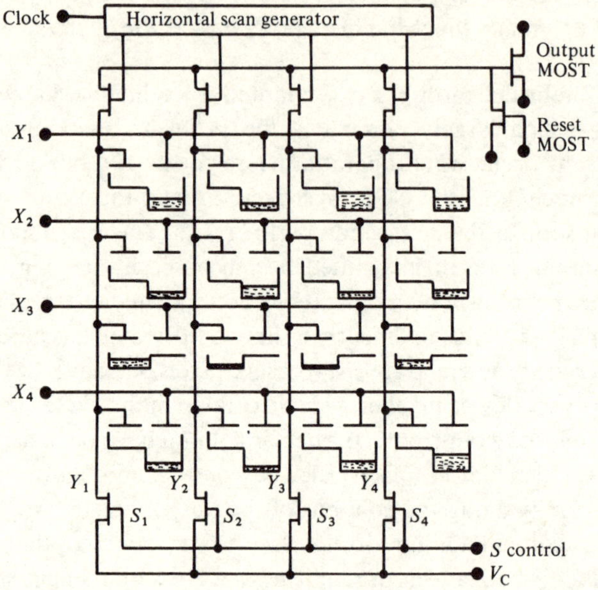

charge while the other electrode is held at the substrate potential. When the electrode potentials are interchanged the charge packet moves to the potential well now created under the other electrode and its presence is detected through capacitive coupling to the electrode. It is important to note that this process is non-destructive so that a continuous monitoring of the integrated charge is possible as it builds up and is a very useful feature when the optimum integration time for a particular astronomical object is not known. One of the electrodes of each cell is connected to the column lines and the other electrode to the row lines of the array. To scan a horizontal row of cells, all row voltage lines are set high and all column voltage lines set to an intermediate level. The column lines are then disconnected from their supply and allowed to float. The voltage is then removed from the row to be scanned whereupon the charge packets in each cell move underneath the column electrodes and the potential of each column changes by an amount equal to the value of the charge packet divided by the particular column capacitance. These changes in potential are then clocked to the output amplifier by the output scanning register. At the end of the final readout all electrodes are set low whereupon the charge packets are forced or injected into the substrate. In the early devices each cell was measured this way and hence the name

CID. The presence of many electrodes connected to each column line and row line results in a high line capacitance which in turn means a lower voltage change per unit charge than for the CCD. The equivalent noise per readout is ten to twenty times higher than for the CCD and is the principal drawback of the device. However, since the readout is non-destructive, multiple readouts can be added to reduce the equivalent noise to be comparable to the CCD. The fact that the charge packets are not passed from cell to cell eliminates charge smearing and the isolated nature of the cells greatly reduces blooming. The CID has been used successfully by several astronomical groups but will probably never seriously compete with the CCD due to the much higher readout noise and the apparent lack of commercial interest in developing devices ideally suited to astronomy.

8.8 Digicon and Reticon diode arrays

The p-n and p-i-n silicon diodes are such linear, wide dynamic range, efficient and simple detectors that several attempts have been made to build arrays of diodes suitable for astronomical imaging. Direct illumination of the diodes would give a very efficient optical detector due to the high quantum efficiency of the silicon diode in the visible and near infrared parts of the spectrum. However the high readout noise associated with the large capacitance of even a reverse biased p-i-n diode requires the use of pre-detector gain to achieve single photoelectron counting capacity. The most astronomically successful use of a diode array is the Digicon linear array developed for use as a spectroscopic detector. Figure 8.14 shows the basic arrangement of a 38 element array used by one research group. Other configurations have been used by other groups with similar results. In this example pre-diode gain is obtained from a magnetically focused image tube which when operated at 20 kV gives an average of 5000 electron–hole pairs in a diode for each primary photoelectron. Even with this large gain and the array cooled to −20 °C, the readout noise would be too high for single photoelectron counting if the diodes were connected directly to the output register, which would be the case for sequential reading of each diode. Each diode is instead connected to its own charge sensitive amplifier, voltage gain amplifier and pulse height discriminator and then the output register. When used in this way not only can single photoelectrons be detected but the sensitivity of all diode channels can be set very nearly equal. The use

Figure 8.14. The Digicon imaging device. (a) Basic mechanical layout showing the magnetically focused intensifier and multi-diode detector. (b) Two representative channels of electronic signal processing circuitry.

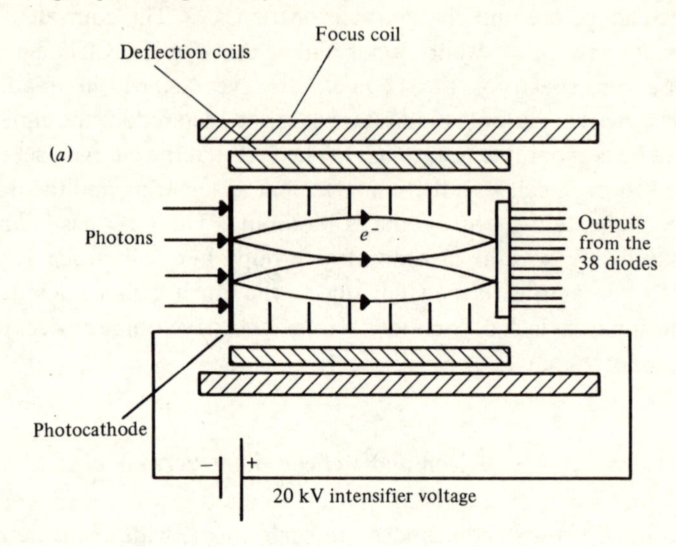

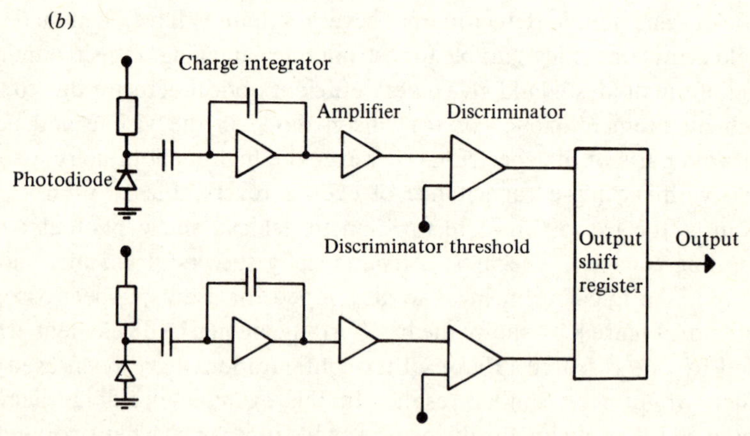

of a magnetically focused image tube allows the image to be effectively stepped across the array to eliminate fixed pattern noise. The discriminator rejects noise pulses from the diodes and charge coupled amplifiers and can be set to give one digital count per primary photoelectron. It has been found possible to set the discriminator so that 80 per cent of photoelectrons are detected which when combined with the quantum efficiency of the photocathode gives an overall detective quantum efficiency peak of about 5 per cent at 400 nm. This is much lower than for CCDs. However the Digicon is a real-time photon detector useful for studying rapidly varying astronomical objects, and has excellent anti-blooming and linearity characteristics. The chief drawbacks are the high cost of the device and the short lifetimes so far encountered by all users. It is thought that bombardment of the diode and amplifier circuits by heavy ions released by the photocathode is the principal destructive mechanism.

An alternative approach to the use of diode arrays is to accept the increased readout noise inherent in direct diode multiplexing but greatly increase the length of the array to some 1000 to 4000 elements. This has been achieved by the Reticon Corporation whose diode array products are usually referred to as Reticon arrays. Figure 8.15 illustrates the essential on-chip circuitry of a 936 × 2 Reticon array. The arrangement of two identical adjacent rows of p-n diodes 375 × 30 μm spaced 30 μm centre-to-centre along the array permits the simultaneous acquisition of two spectra, for example a star and the background sky. Each photodiode is connected to one of four output lines by means of a transistor switch driven from a two-phase address shift register. For each row, the odd and even diodes have separate output lines. The two odd lines and two even lines are each connected together and to a charge sensitive amplifier. This amplifier is connected to a sample and hold module which temporarily stores one diode's integrated charge until the analogue switch can pass the charge on to the remainder of the signal processing circuitry. Switching off both address registers disconnects all diodes so that long term integra-

Figure 8.15. Electronic layout of a Reticon array with the photodiodes and circuitry arranged in two parallel rows.

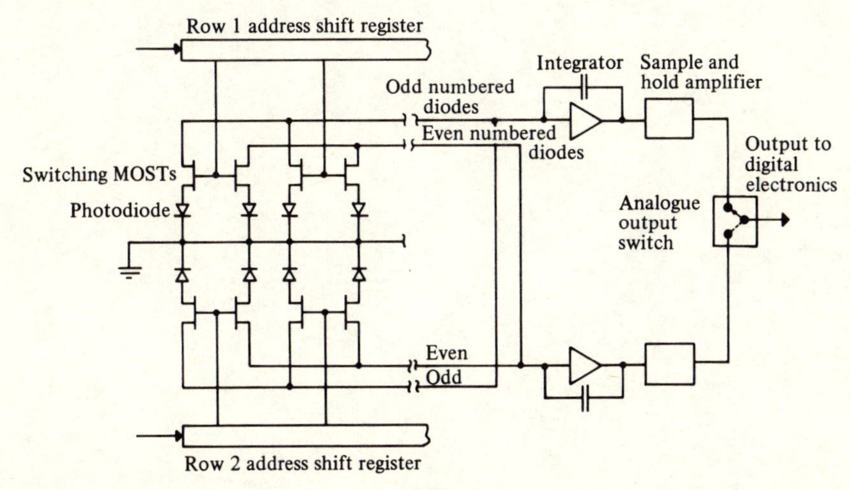

tions are possible. It is possible to use the Reticon array for direct imaging but, as mentioned earlier, practical astronomical usage requires pre-detection gain. This is usually achieved by cascaded magnetically or electrostatically focused image tubes optically coupled to the cooled array by either lenses or fibre optics. The principal advantage of the Reticon array is its commercial availability and, when preceded by an optical gain of at least 10 000, its ability to detect single photoelectrons. This large optical gain requires such a high voltage across the image intensifier that problems such as corona and surface leakage can occur unless care is taken when potting the high voltage components. The quantum efficiency of the overall system is that of the primary photocathode and will usually be 10 per cent or less. While not as sensitive as the Digicon or CCD, the Reticon array is a useful astronomical detector on which work is continuing.

References for Chapter 8

Beynon, J.D.E. & Lamb, D.R., 1980. *Charge-Coupled Devices and Their Applications*, McGraw-Hill, London.

Howes, M.J. & Morgan, D.V. (eds.), 1979. *Charge-Coupled Devices and Systems*, Wiley, Chichester.

International Conference on Technology and Applications of Charge Coupled Devices, 1974. University of Edinburgh.

Melen, R. & Buss, D. (eds.), 1976. *Charge-Coupled Devices: Technology and Applications*, IEEE Press, New York.

Index

Index

Index

Index